HOTEL \ OFFICE \ CLUB 酒店\办公\会所

GOLD MANTIS DESIGNING

金螳螂设计

《金螳螂设计》编委会 编
傅克玲 高 健 译

辽宁科学技术出版社

本书编委会

出版说明 PREFACE

金螳螂建筑装饰股份有限公司成立于1993年，是具有施工一级、设计甲级资质的企业，是集室内装饰、幕墙、家具、景观、艺术品、智能化安装等相关配套产业为一体的专业施工企业，是全国装饰行业的首家上市公司、首家拥有中国驰名商标的装饰企业，至今已连续五年荣获中国建筑装饰百强企业第一名。

“金螳螂”是从事酒店、商场、办公楼、娱乐场所、医院、体育场馆等各类公共建筑装饰设计和施工的专业公司，拥有600多人的国内最大室内设计师团队及近200多名注册建造师、一、二级项目经理队伍，在全国20多个大中城市承接了大批中高档工程。秉承“实力缔造完美空间”的理念，近年来先后完成了北京人民大会堂江苏厅、首都博物馆、中国丝绸博物馆、第28界世界遗产大会主会场、常州富都香格里拉大酒店、苏州东山国宾馆、山西晋祠国宾馆、中国网通会所、北京友谊医院、南京奥林匹克体育中心等大型项目的设计及施工，得到了客户的一致好评。

本书汇集了金螳螂建筑装饰股份有限公司近年来优秀的设计案例，包括酒店、办公和会所共计一百多个项目，采用设计说明、平面布置图和精彩效果图相结合等方式，诠释了金螳螂建筑装饰股份有限公司的设计宗旨，是不可多得的室内设计专业图书，既可供读者开拓思路，又可直接借鉴使用。

Gold Mantis Construction & Decoration Co., Ltd. was founded in 1993. Now, it has the qualification of first-class construction and design for building decoration and is a specialized construction enterprise which involves in such related industries as interior decoration, curtain wall, furniture, landscape, artworks and intelligent installation. It is the first stock company in the decoration industry of China and the first decoration company that has been awarded “Chinese Famous Trademark”. The company has been awarded as Number One in “Top Hundred Building Decoration Enterprises in China” for successive 5 years.

Gold Mantis Construction & Decoration Co., Ltd. is specializing in designing and decorating of various public buildings as well as enterprise’s buildings. It owns over 600 interior designers who form the largest interior design team in China. It also has more than 200 certified constructors and a level A and Level B project management team. In more than 20 large and medium-sized cities, the company has undertaken a large number of mid or high level projects. With the philosophy of creating perfect space, the company has recently finished the design /construction of Jiangsu Office in the Great Hall of the People in Peking, Capital Museum, China Silk Museum, Main Venue of the 28th World Heritage Conference, Changzhou Shangri-la Fudu Hotel, Suzhou Dongshan Hotel, shanxi Jinci Hotel, China Netcom Clubhouse, Beijing Friendship Hospital, Nanjing Olympic Gym Center.

This book compiles all the excellent designs conducted by the company in the past few years, including more than 100 designs of hotels, offices and club houses, etc. Combining brief explanations, layout plans and vivid effect drawings, this book is a perfect explanation of what the company is all about. It is a valuable book of interior designing that can not only help the readers expand their minds, but also can be used for reference directly.

CONTENTS 目录

酒店部分
HOTEL PART

办公部分
OFFICE PART

会所部分
CLUB PART

昆山清风明月酒店

KUNSHAN QINGFENG MINGYUE HOTEL

设计负责人：席行千　Architect: Xi Xingqian

设计参加人：张辉、张中斌、朱庆新　Collaborator: Zhang Hui, Zhang Zhongbin, Zhu Qingxin

建筑面积：3500㎡　Building Size: 3500m²

工程造价：800万　Construction Cost: RMB 8 million

获奖情况：荣获"最佳包厢设计作品"奖　Awards: Best Designing Works for Private Rooms

"清风明月本无价，近水远山皆有情"。苏州沧浪亭里的这两句著名诗对就是清风明月酒家店名的由来。坐落于江南名城、昆曲发祥地昆山的清风明月大酒店，紧邻昆山市政府大楼，地理位置得天独厚，周边景观疏朗开阔。酒店面积3600平方米，可同时提供1180个餐位。

酒店装饰装潢亦素亦华，亦中亦西，亦古亦今，设计中处处显现出对环境设计精细、独到的追求。灯具在清风明月也是一道风景：或方，或圆，或长，或短，或伞形，或腰形。除了照明功能，更多的是一种环境的点缀,情绪的渲染，温馨中透着恬适、甜蜜，催人遐想，令人流连忘返。走出风格别致的包厢，呈现在你眼前的是疏朗空阔、气派不凡的大厅。每一处摆设和布置，都充分体现精细雕琢的设计理念。

夜晚的"清风明月"，更是别具景象。灯光打在木制的装饰墙壁上，墙壁反射回的柔和光线，洒落在一件件精美的艺术品上，此刻，你也许会想起流光溢彩四个字。——对，只有流光溢彩四个字，才配描写这样的视觉感受。

The hotel's name originates from the famous couplet in Canglang Pavilion, "Qingfen Mingyue Ben Wujia, Jinshui Yuanshan Jie YouQing". That means the refreshing wind and shinning moon bring much favor to people. Which is priceless. Likewise, seeing through the window, the far-away mountain and nearby water all hold a deep love to the people. Located in the famous city of the South Yangtze River, Kunshan, the birthplace of Kun opera, Kunshan Qingfeng Mingyue Hotel stands side by side with Kunshan municipal government buildings, with its unique geographical advantage. The surroundings are open and beautiful. The hotel is 3600m² in size and can accommodate 1180 dining seats at the same time.

The decoration of the hotel is both simple and luxurious, Chinese and Western, ancient and up-to-date. The design shows clearly its meticulous and unique pursuit of environment perfection. Lamps and lanterns are also beautiful sceneries in the hotel: square or round, long or short, in the shape of an umbrella or a waist. Apart from their lighting function, they are more of the decoration of environment, the playing up of affection —— showing quietness, comfort and sweetness with the tender feelings, which can easily arouse your imagination, make you linger about. Stepping out of the private rooms of unique style, what is spreading before you is the spacious and splendid lobby. Each ornament and arrangement indicates the designing idea of pursuing perfection.

The hotel takes on a distinctive view at night. When the lamps and lanterns cast light on the wood-decorated wall, the soft light the wall reflects may spread on the exquisite works of art. Then what you will think of is the four Chinese characters —— liu guang yi cai, meaning the light seems as if streaming, while the colors are quite beautiful. Yes, only the four Chinese characters deserve such a visual sense.

篤信

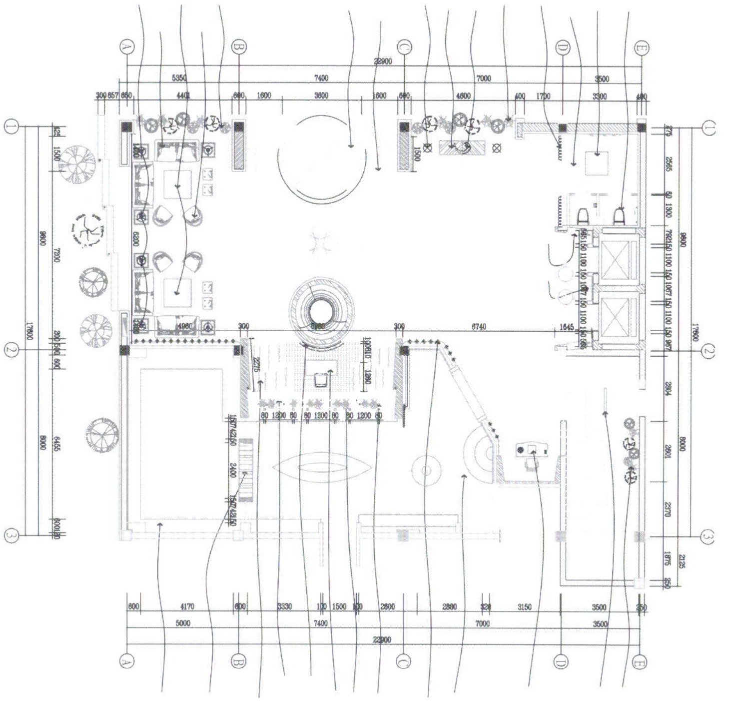

科技新天地酒店

SCIENCE NEW WORLD HOTEL

设计负责人：李泉 Architect: Li Quan

设计参加人：席行千、朱志霞、徐杰、尹胜利 Collaborator: Xi Xingqian, Zhu Zhixia, Xu Jie, Yin Shengli

建筑面积：109 600m^2 Building Size: 109 600m^2

工程造价：5700万元 Construction Cost: RMB 57 million

本案位于苏州机场路以南、通园路以东的湖边，占地面积36 800.36平方米，是集研发、办公、酒店式公寓、商业综合配套等多种功能为一体的组团建筑群，总建筑面积151 874.09平方米。

该酒店的室内设计在风格上延续建筑语言，同时又有所突破。用纯净、简洁、明快、时代感极强的设计手法来定位空间气质，运用美轮美奂的光环境来营造空间氛围，给建筑赋予新的生命力。强调室内环境的默契与融合，兼顾空间序列的进深层次。注重空间尺度的把握，色彩的运用，材料质感的表达，使整体设计风格完整而统一。

The hotel is located on the Lakeside to the south of Suzhou airport road, and the east of Tongyuan road, with an occupation area of 36 800.36m^2. It is the complex of buildings for research and development, offices, hotel-type apartments and commercial centers. Its total floor space is 151 874.09 m^2.

The interior designing style is retaining the architecture art and has some breakthrough as well. It is using pure, brief and lively techniques of distinct characteristics of our times to orientate its space quality, using its splendid lights to create space atmospheres, to give the building a new vitality. It gives importance to the mixture and harmony of the interior environment, also gives consideration to the different level of space order. It values the perfect control of space size, the use of colors and the express of material qualities, making the whole designing style integrated and unified.

舟山财富大饭店

ZHOUSHAN FORTUNE HOTEL

设计负责人：童向东　Architect: Tong Xiangdong
设计参加人：周斌、程均清　Collaborator: Zhou Bin, Cheng Junqing
建筑面积：30 000㎡　Building Size: 30 000㎡

将自然、生命的主体元素以变形、错位、符号化等多种创作手法加以提炼，以抽象和象征手法引发人们的想象和联想。进入大堂，透过云石组成的大小不同的圆，以体现一种“禅”意，给人以丰富的想象；莲花形的地面石材拼花与顶棚造型相呼应，吊顶中间悬着赋有雕塑感的金色造型挂饰，似帆似鱼翅，体现一种吉祥和财富的象征；灯笼造型的吊灯和反映舟山特色的沙雕造型，又让人感受到中国情怀。

The design deploys various techniques such as transformation, displacement and symbolization, etc. to refine the subjective elements of the nature and life, and the abstract and symbolic techniques are used to arouse the imagination and association. Inside the lobby, different sizes of circles made of translucent marble contain a Chan message, bringing profuse imagination. The lotus-pattern stone mosaic floor matches well with the roof designing. Hanging in the center of the ceiling, the golden pendant of sculpture feature looks like a sail or a shark's fin, token of good luck and wealth. The lantern-shaped pendant lamp and the sand-sculpture of Zhoushan feature remind guests of its Chinese feelings.

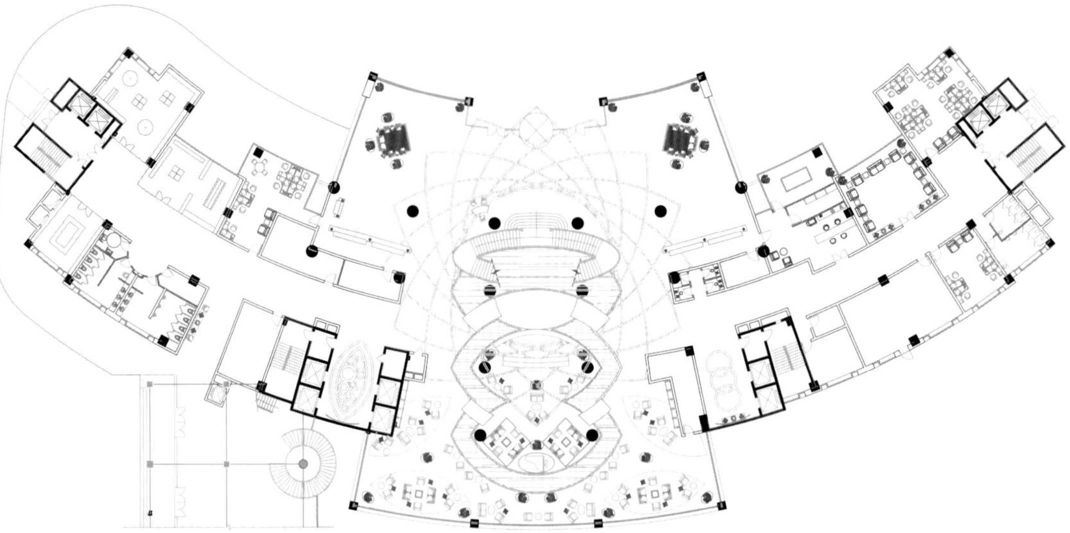

象山国际世贸酒店

XIANGSHAN WORLD TRADE CENTER HOTEL

设计负责人：郭轲蔚　Architect: Guo Kewei

设计参加人：司马东兰　Collaborator: Sima Donglan

建筑面积：28 000㎡　Building size: 28 000㎡

工程造价：1.5亿元　Construction Cost: RMB 150 million

获奖情况：荣获“最佳酒店作品”奖　Awards: Best Works for Hotel Design

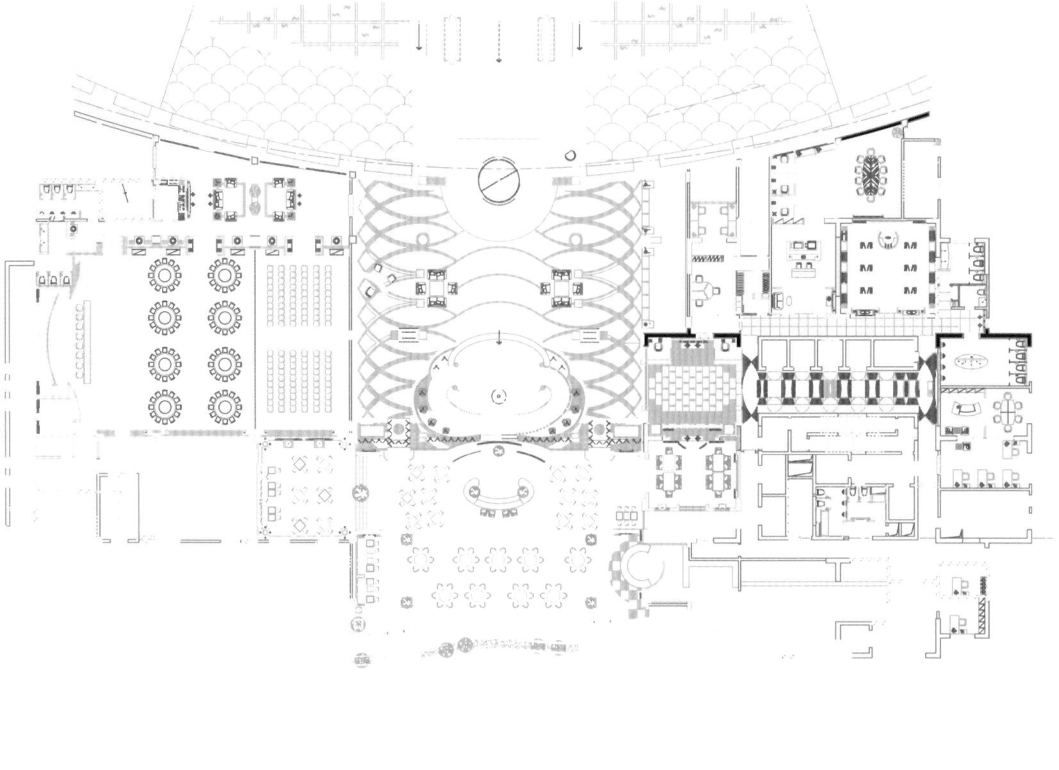

扬州观邸酒店

YANGZHOU GUANDI HOTEL

设计负责人：丁飞 Architect: Ding Fei

设计参加人：贝炜瑾，鲁晴芋，张颖，钱丽芬，张俊 Collaborator: Bei Weijin, Lu Qingyu, Zhang Ying, Qian Lifen, Zhang Jun

建筑面积：2500㎡ Building Size: 2500m²

工程造价：400万元 Construction Cost: RMB ￥4million

"天下三分明月夜，二分无耐在扬州"。

观邸酒店坐落在诗情画意的古运河畔。伴着皎洁的明月、荡漾的河水，品着新概念极致菜肴，耳中仿佛全是清脆的渔鼓，悠扬的歌声。在这里时间仿佛凝止，人生的忧郁、烦恼均融入了清清的河水，淡淡的消散。

观邸是一座怀旧风格的建筑，它取决于"坐观天下"之意，借鉴于旧上海外滩建筑及民国建筑的风格，立面细腻不张扬。遥望观邸，仿佛又聆听到夜上海飘渺的歌声。穿过仿古的石库门洞，便进入了一个幽雅，尊贵的殿堂，室内装饰中西合璧，古今交融。观邸的整体设计采用了新古典主义的理念。而在平面布局中，则融入了姑苏古街坊之意境。唐朝兴盛时曾在苏州划分了六十古坊，而在这里借苏州古坊之名，露出了浓浓的怀旧之意，从入口至最后一个VIP包厢，以姑苏繁华图予以铺展，并用姑苏繁华图点睛，仿佛来此就餐，就是对一段记忆的回顾。

If the moonlight can be divided into three parts, two parts are casting onto Yangzhou, while the other places can only share the last part.

Guandi hotel is located beside the ancient and poetic Jinghang Cannel. Accompanied by the bright moonlight, the waving water, you can enjoy the most splendid food, and you will feel as if the local percussion instrument is playing in your ear, and the melodious songs are haunting. Here you will sense as though the time stopped, and all the melancholy and worries in life seem to be blending into the clear river, dispersing gradually.

Guandi is a building of nostalgic style. Guandi means "looking on peacefully", which is an attitude towards life. It borrows the building styles of old Shanghai bund and the Republic of China. The vertical plane is exquisite but not showing off. Looking at Guandi from a distance, you are intended to feel as if you were enjoying the visionary songs from the Night Shanghai. Passing through the ancient style stone gateway, you will come into the quiet and lobby, which is well decorated in the combination of Chinese and Western style, in the harmony of ancient and modern techniques. The whole design adopts the neoclassical notions. As for the plane arrangement, it also blends with the artistic conception of Gusu ancient workshops. During the glorious Tang Dynasty Suzhou was once divided into sixty workshops. Here by borrowing the names of Gusu ancient workshops, it reveals strongly the feeling of recalling the past times. From the entrance to the last VIP private room, it can be seen that the flourishing picture of Gusu is the clue and also the striking key point of the arrangement. Dining here seems like a recalling of a beautiful memory of the past.

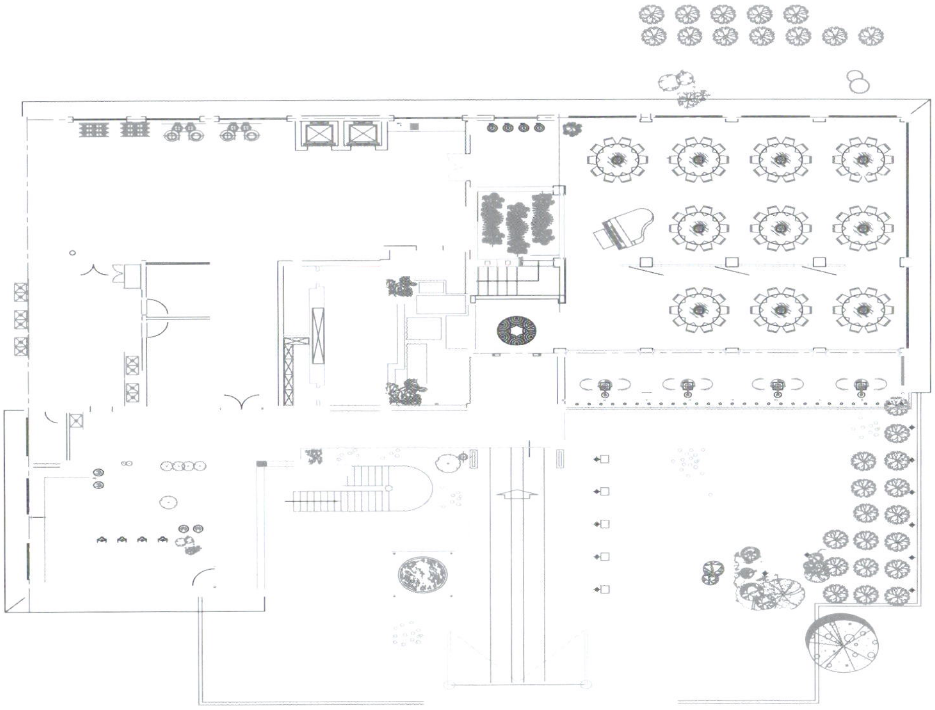

雅都大酒店

YADU HOTEL

设计负责人:周奕君　Architect: Zhou Yijun

设计参加人:席行千、李泉、朱志霞、徐杰　Collaborator: Xi Xingqian, Li Quan, Zhu Zhixia, Xujie

建筑面积:22 048m^2　Building Size: 22 048m^2

工程造价:2500万元　Construction Cost: RMB 25 million

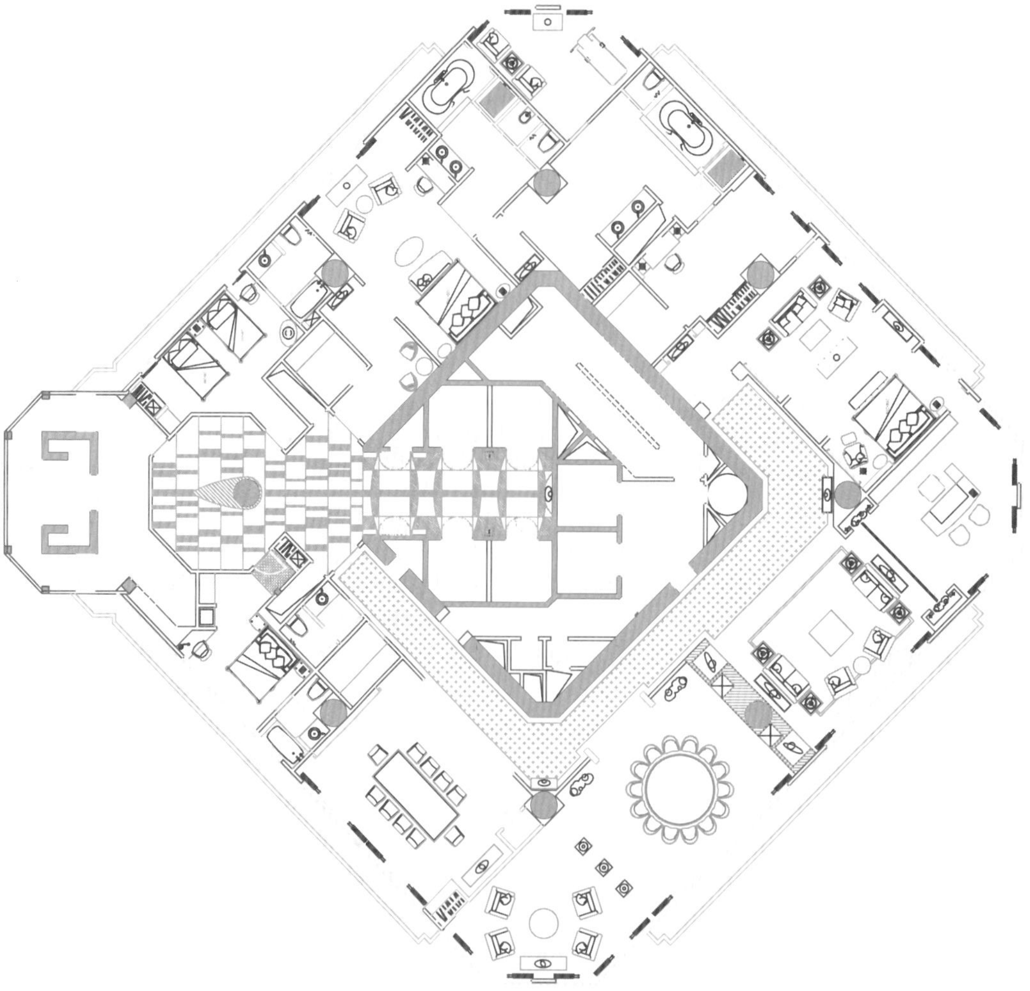

南京紫金山庄

THE PURPLE PALACE NANJING

设计负责人:高超一、范健中 Architect: Gao Chaoyi, Fan Jianzhong
设计参加人:沈辉、朱文奇、刘云龙、袁珏梅、莫奇敏 Collaborator: Shen Hui, Zhu Wenqi, Liu Yunlong, Yuan Juemei, Mo Qimin
建筑面积:22 000m² Building Size: 22 000m²

本案处于南京东郊钟山之麓的黄马水库。绿树掩映，清波荡漾，古迹遍布，山清水秀，地理位置十分优越。设计宗旨围绕体现项目设计的功能和要求。

帝王的贵气：高规格的政府接待和高级商务活动中心；
江南的文气：要求体现尊贵、庄严、完美、安全；
地域的灵气：舒适的气氛。
设计师在策划本案主楼时根据主楼的功能需要，着力体现“虎踞龙蟠”的帝王之气。

本案室内设计强调了“红”与“黑”的对话，又以“白”和“金”为协调色。红为中国传统的朱砂红，黑为中国黑，白为汉白玉，金为纯金箔。

Beside Huangma Reservoir at the foot of Zhongshan Mountain in the east suburb of Nanjing, the Purple Palace Nanjing is ideally situated. Green trees, rippling lake water, and historic sites all over constitute its fantastic scenery. So the design is aiming at:

Regal Elegance — high level of government reception center and business center.
South Yangtze Cultural Disposition — embodying honor, dignity, perfection and security.
Regional Inspiration — a comfortable atmospheres.
To meet the need of the main building’s function, the designer is trying to embody the royal elegance.

The interior design emphasizes the contrast of red and black, and chooses white and golden as the coordinate colors. Red is the traditional Chinese cinnabar. Black means the Chinese black. White refers to white marble. Golden stands for pure gold leaf.

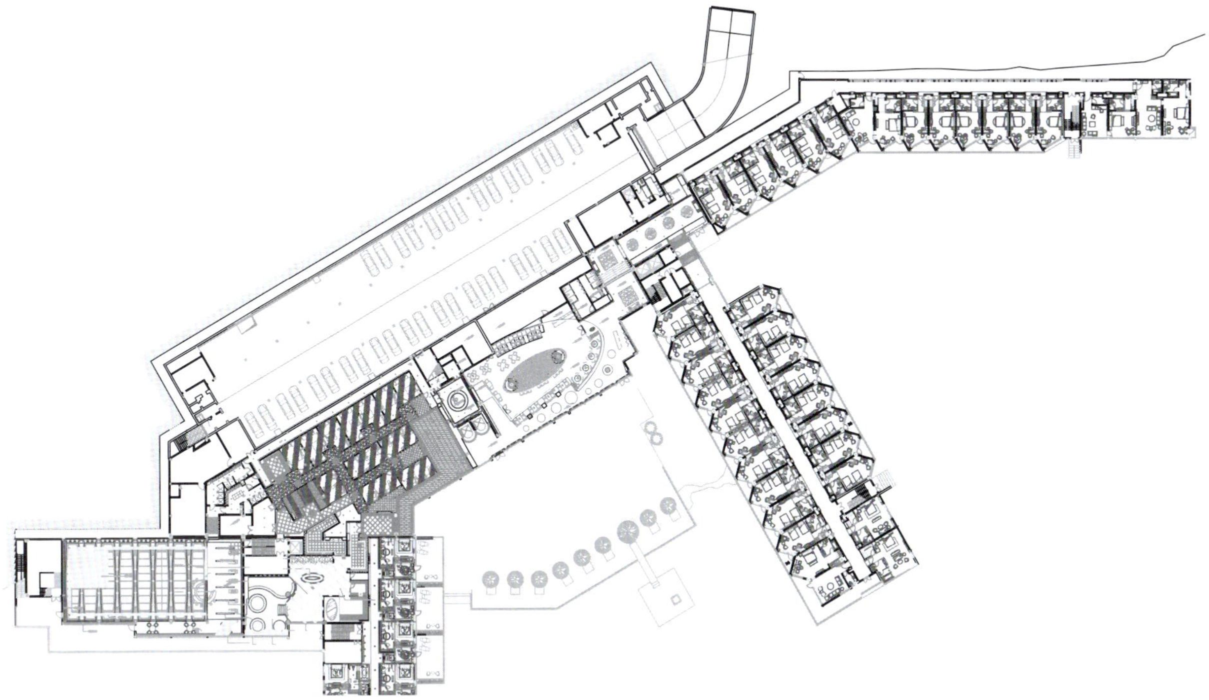

Welcome

无锡香梅国际大酒店

WUXI XIANGMEI INTERNATIONAL HOTEL

设计负责人：李勇强　Architect: Li Yongqiang

设计参加人：钱钢、林硕、邱瑜、方婷、浦晨　Collaborator: Qian Gang, Lin Shuo, Qiu Yu, Fang Ting, Pu Chen

建筑面积：20 000m²　Building Size: 20 000m²

工程造价：50 000万元　Construction Cost: RMB 500 million

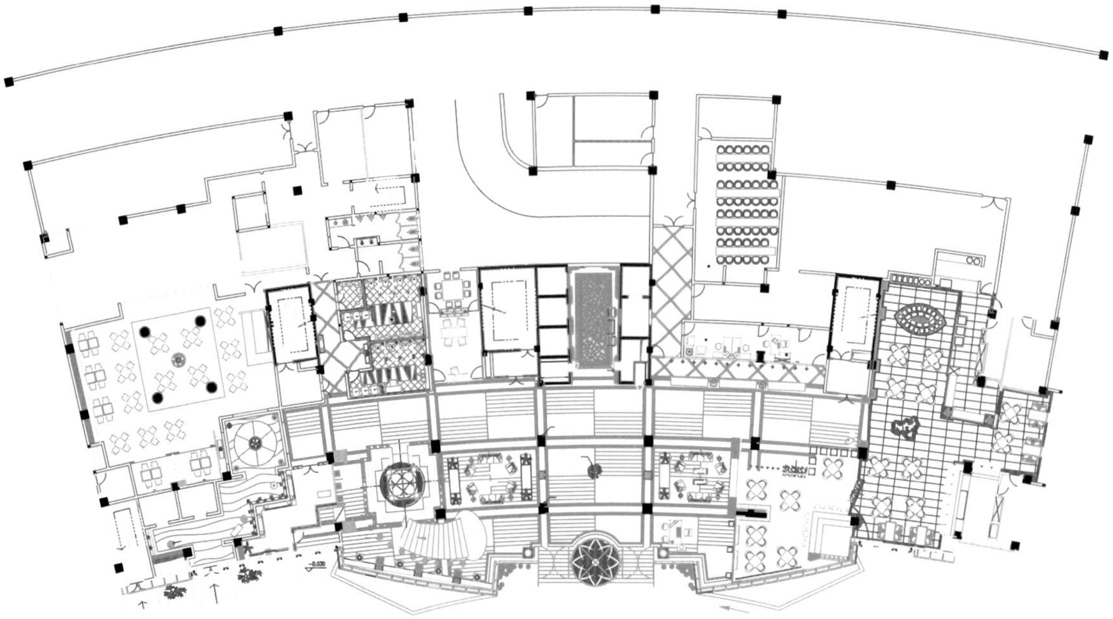

东山宾馆三期

DONGSHAN HOTEL (PHASE III)

设计负责人:朱伟刚、姜亚洲　Architect: Zhu Weigang Jiang Yazhou

设计参加人:陈长斌、王宇新、杨锋、刘庆惠、陆屹、黄懿、孙静

Collaborator: Chen Changbin, Wang Yuxin, Yang feng, Liu Qinghui, Lu Yi ,Huang Yi, Sun Jing

建筑面积:20 000m^2　Building Size: 20 000m^2

蔚蓝海岸大酒店

BLUE COAST HOTEL

设计负责人:朱伟刚　Architect: Zhu Weigang

设计参加人:魏永康、吴捷林、杨锋、刘庆惠　Collaborator: Wei Yongkang, Wu Jielin, Yang Feng, Liu Qinghui

建筑面积:10 000m²　Building Size: 10 000m²

工程造价: 2000万元　Construction Cost: RMB 20 million

地处有着龙城之美誉的常州的蔚蓝海岸大酒店，是一座按四星级标准建造的酒店。

为使该酒店无论是在风格特征或是在营运功能及服务流线等方面都达到良好的效果，将该酒店的原建筑空间进行了整合。为了使大堂更具豪华感，在一、二层之间增设了共享空间，同时为减少二层餐饮区域用餐人群的嘈杂声对一层大堂所产生的不良效果，故将二层沿栏杆处全部封闭。

在总统房的设计中，根据业主要求，同时也是结合经营实际需要，将总统房的多处功能空间都设置成可独立出售的形式，如随从房、夫人房、棋牌室等。如此的设计既满足了星级评定的要求，同时也为业主提供了灵活多变的经营空间。

总之，功能是首要的，如果功能没有了，那么形式也将成为无本之木，无源之水。

Located in Changzhou, which enjoys the fame as the city of dragon, Blue Coast Hotel is constructed according to four-star standard.

In order to achieve the best effect of style and feature, operational functions as well as serving process for the hotel, its original architectural space has been integrated. To make it more sumptuous, the shared space between the first and second floor is added to the lobby, meanwhile the second floor is enclosed from along the railing, so that the noisy effect on the lobby from the dining area of the second floor can be reduced.

For the design of Presidential suites, many of their functional spaces, for example, attendant rooms, madam rooms, chess & cards rooms are designed to be in the separately salable forms, considering proprietors' demands and the actual needs of operation. Such designs can meet the need of star-grading for the hotel and supply proprietors with a flexible operational space as well.

In all, function is of vital importance. Without its function, forms will be a tree without roots, a river without a source.

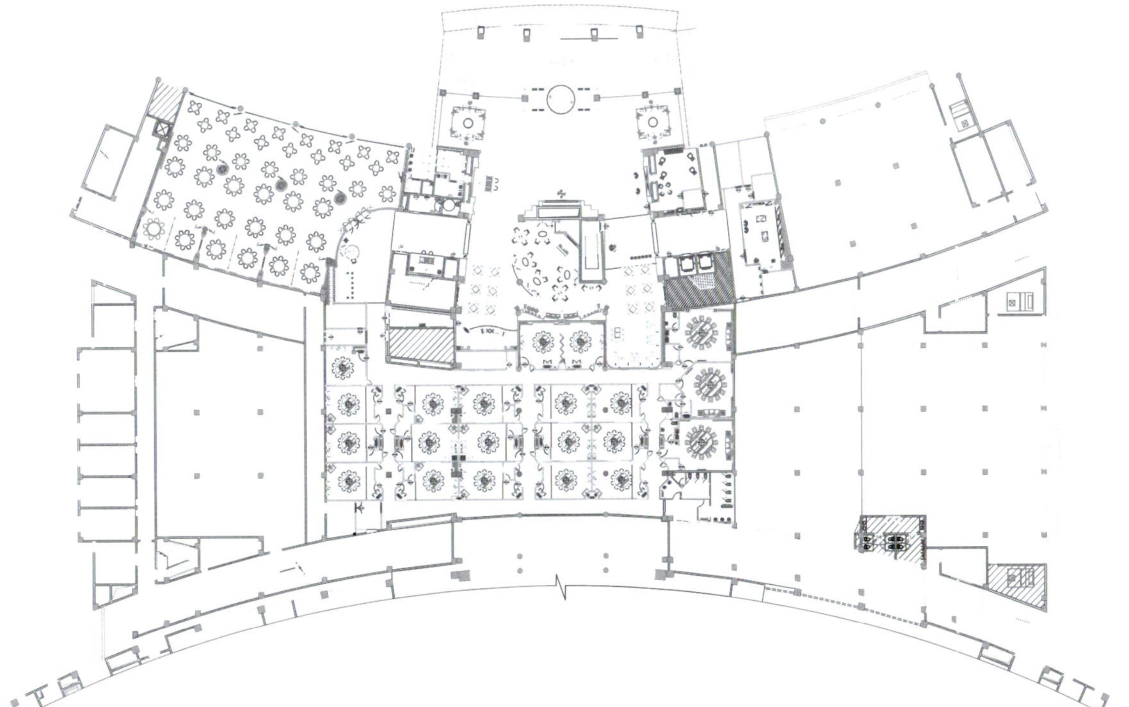

黄山国际大酒店

MOUNT HUANGSHAN INTERNATIONAL HOTEL

设计负责人:季震宇 Architect: Ji Zhenyu

设计参加人:童超、梁爱勇、谢燕、李娴 Collaborator: Tong Chao, Liang Aiyong, Xie Yan, Li Xian

建筑面积:14 500m² Building Size: 14 500m²

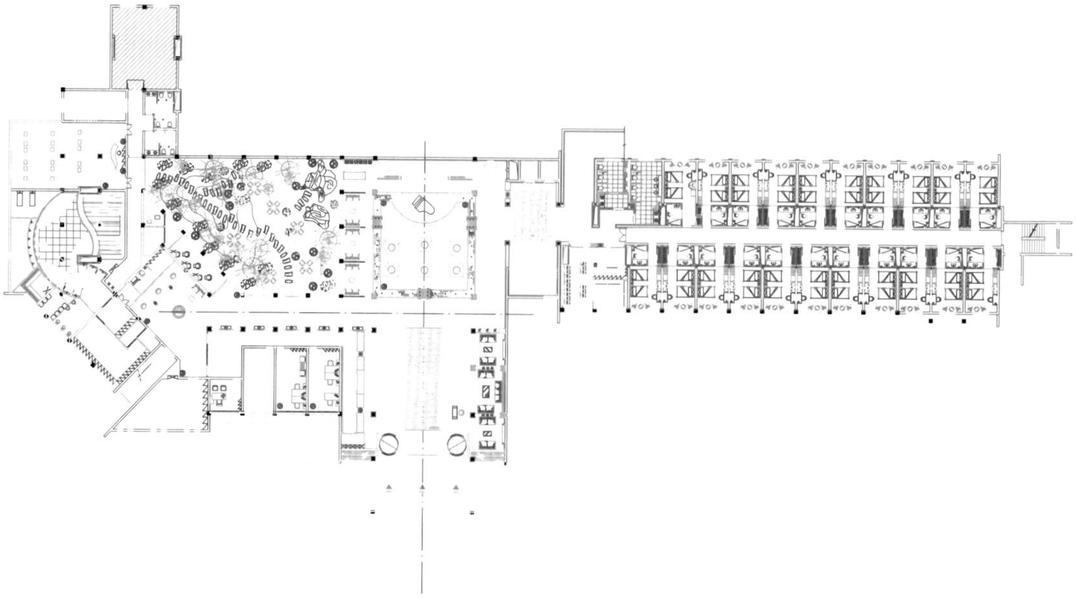

广州南美大酒店

GUANGZHOU NANMEI HOTEL

设计负责人：高超一、李朝中　Architect: Gao Chaoyi, Li Chaozhong
设计参加人：夏芹　Collaborator: Xia Qin
建筑面积：28 000㎡　Building Size: 28 000m^2

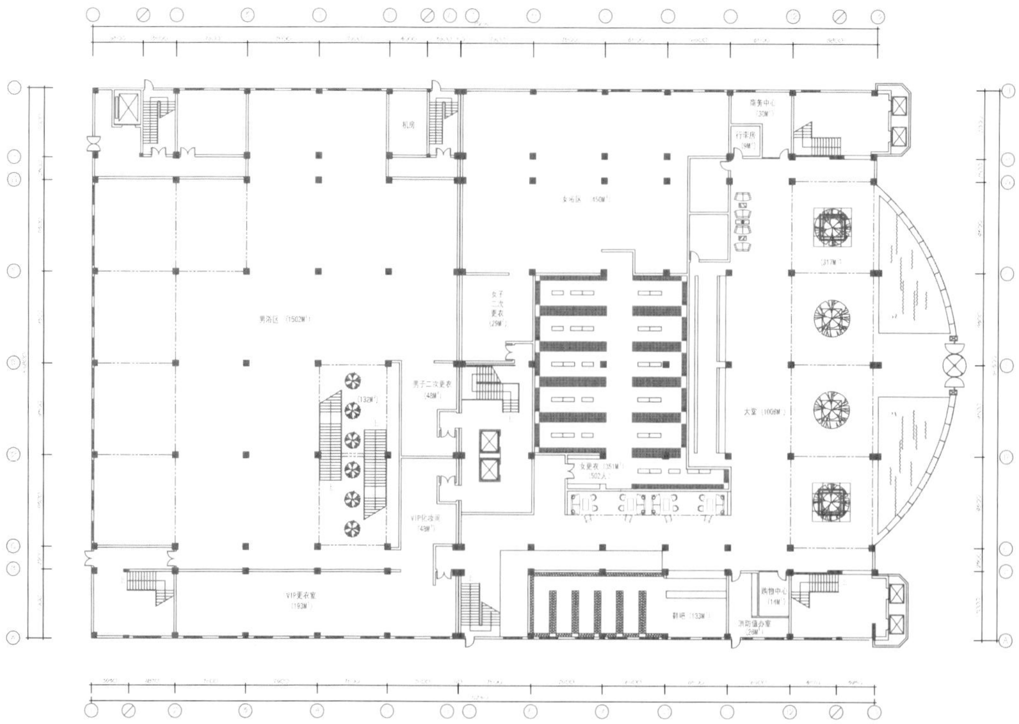

同里湖度假村

TONGLIHU RESORT VILLAGE

设计负责人：马晓星、倪苏宁、张恒祥 Architect: Ma Xiaoxing, Ni Suning, Zhang Hengxiang
设计参加人：刘治平、梁明阳 Collaborator: Liu Zhiping, Liang Mingyang
建筑面积：10 176m^2 Building Size: 10 176m^2

风格定位

同里湖度假村具有独特的优势。首先，其十年历史所铸就的品牌效应在苏州、上海及周边城市有口皆碑；其次，其庭院式的体量特征，十年历程的园内树木生长及其优良地理环境都可成为其将来绝佳的卖点。

如何发挥度假村的优势，造就出一片截然不同的又具独特环境气氛和空间形态特征就成为了设计策划的主要着眼点。

考虑到与同里湖度假村相邻的同里湖大饭店，其风格、造型、材料的使用均较中性，空间表情仍有严肃的一面，度假商务兼娱乐。因而，设计将同里湖度假村的风格定位在太平洋亚热带风格，同时追求一种田园牧歌的生活情趣，营造一种可使人全身心放松，室内与室外自然与人工完全融为一体的环境氛围。

处理手法

以大体块来统一室内造型，追求一种自然古朴的特征；多使用一些容易使人与自然产生联想的天然木材、石材，并尽量保留材料天然肌理，同时为保持度假村的档次，我们仍使用了大量现代经典材料，使游客在最大程度地体验休闲度假环境气氛的同时仍能体验到现代的精神气息。

Style Orientation

Tonglihu Resort Village enjoys its unique advantage. Firstly, its trade-mark effect with ten years of cultivating is universally acclaimed in Suzou, Shanghai and nearby cities. Secondly, its courtyard feature, including ten years of trees growing in the yard and the perfect surroundings, makes the future best selling point. So how to make the most of the advantage to create a completely different and special environment atmosphere and space feature becomes the key point of the design and plan.

Considering the nearby Tonglihu Hotel, a hotel for vacation, business and entertainment, which takes on the neutral feature in its style, shape and material choose, showing somewhat serious aspect of the hotel. Tonglihu Resort Village is designed to be of Pacific subtropical style, with an idyllic pastoral life temperament. The design is also aimed at creating a relaxing, harmonious atmosphere, to make indoor and outdoor scene, natural and artificial things blend perfectly.

Designing Style

Big blocks are used to unify the interior design and to pursuit a natural and primitive simplicity. A lot of natural timbers and stones with their best natural texture are used, which can easily arouse the imagination with nature. Meanwhile, in order to keep up the high grade of the Resort Village, the design also uses a great deal of classic modern materials. Thus the guests can not only fully enjoy the relaxing and vocational atmosphere, but also can experience the modern air.

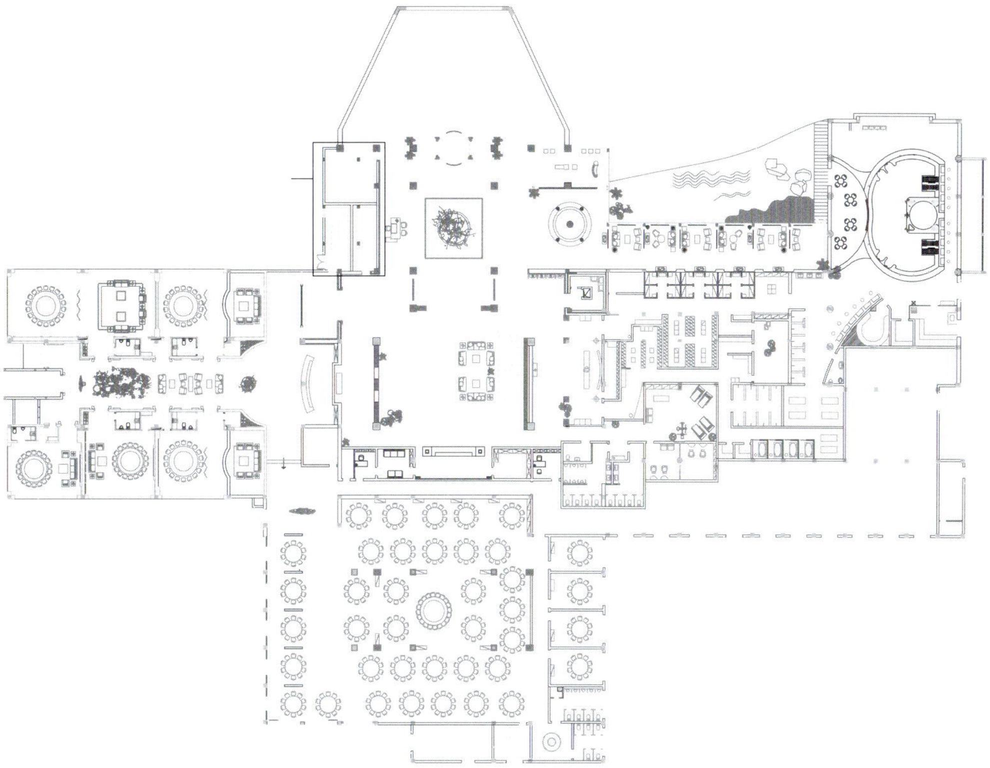

南通综艺大酒店

NANTONG ZONGYI HOTEL

设计负责人：马晓星、倪苏宁　Architect: Ma Xiaoxing, Ni Suning
设计参加人：惠炜、吴再想、苏小丰　Collaborator: Hui Wei, Wu Zaixiang, Su Xiaofeng
建筑面积：22 000m²　Building Size: 22 000m²

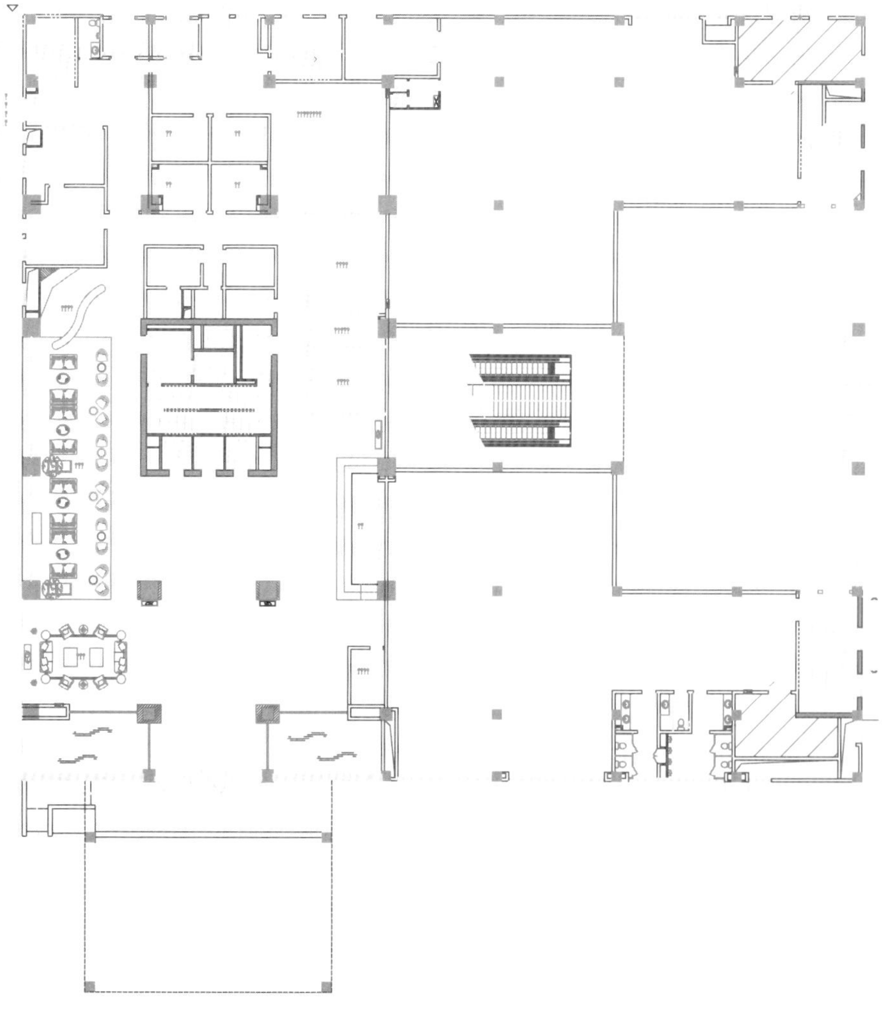

耀江大酒店

YAOJIANG HOTEL

设计负责人:姜樱　Architect: Jiang Ying
设计参加人:潘芳、翁其恂、黄健　Collaborator: Pan Fang, Weng Qixun, Huang Jian
建筑面积:60 000m² Building Size: 60 000m²
工程造价:3000万元　Construction Cost: RMB 30 million

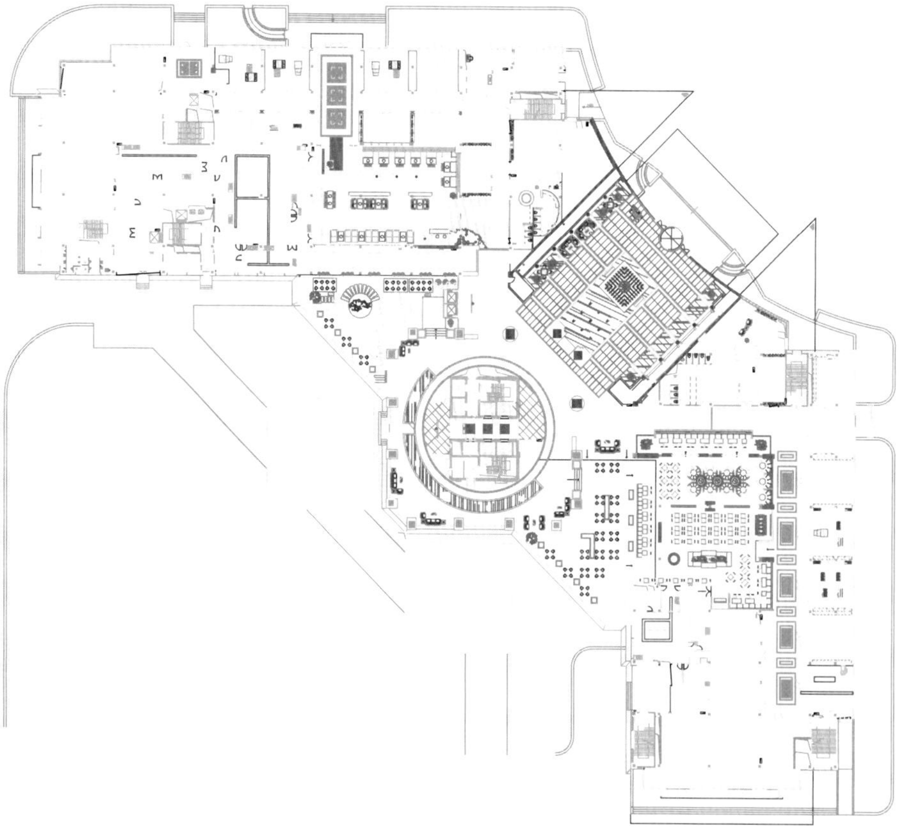

扬州蓝天宾馆

YANGZHOU LANTIAN HOTEL

设计负责人：梁戏、马云飞　Architect: Liang Xiao, Ma Yunfei
设计参加人：刘克勤、沈洪俊　Collaborator: Liu Keqin, Shen Hongjun
建筑面积：1200m²　Building Size: 1200m²
工程造价：600万元　Construction Cost: RMB 6 million

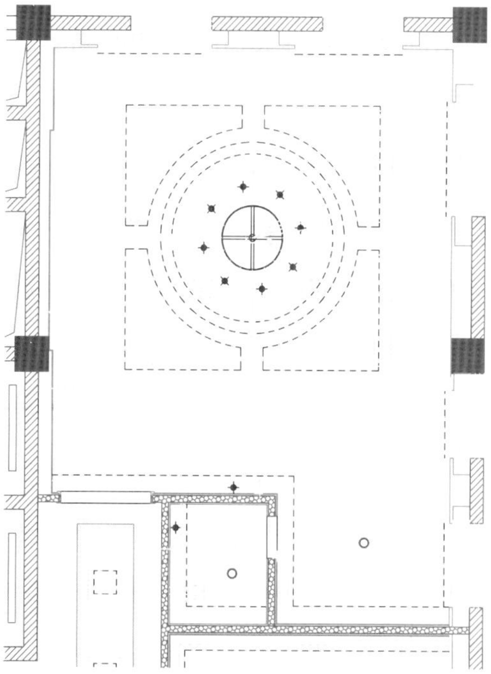

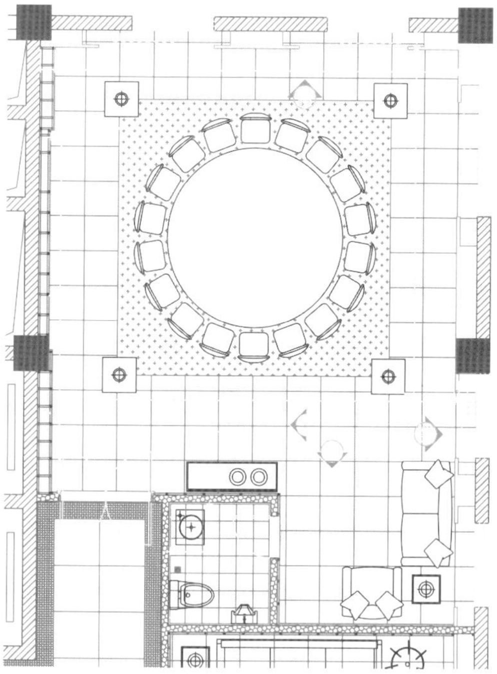

青岛广业锦江大酒店

QINGDAO GUANGYE JINJIANG HOTEL

设计负责人：姜樱　Architect: Jiang Ying

设计参加人：段秀丽、潘芳、丁黎敏　Collaborator: Duan Xiuli, Pan Fang, Ding Limin

建筑面积：25 000㎡　Building Size: 25 000m^2

工程造价：3000万元　Construction Cost: RMB 30 million

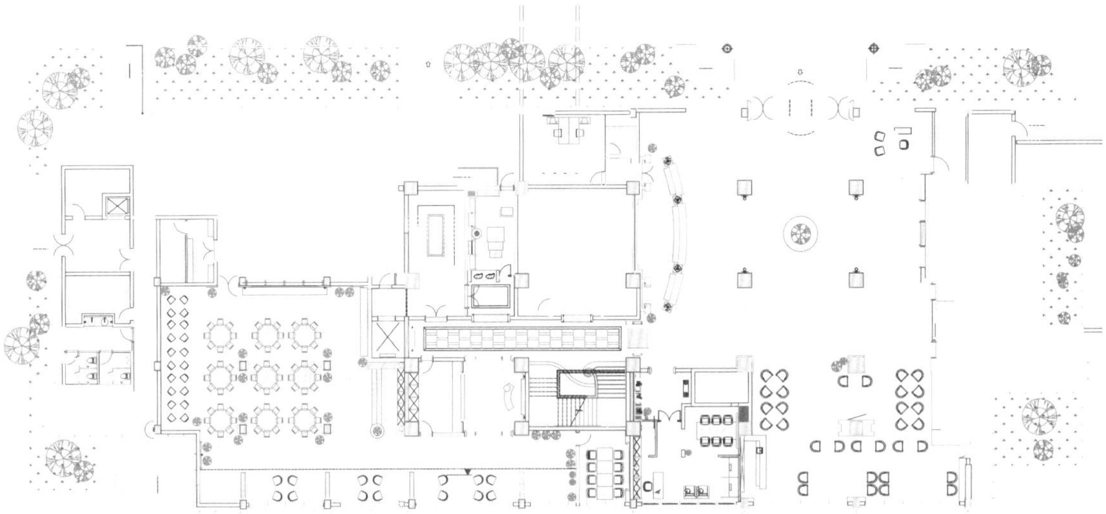

苏州西山银湖山庄

XISHAN SILVER LAKE HOTEL, SUZHOU

设计负责人：包岳良　Architect: Bao Yueliang

设计参加人：黄强、王昕　Collaborator: Huang Qiang, Wang Xin

建筑面积：11 000m^2　Building Size: 11 000m^2

工程造价：1000万元　Construction Cost: RMB 10 million

获奖情况：锦绣江南一等奖　Awards: Jin Xiu Jiang Nan First Prize

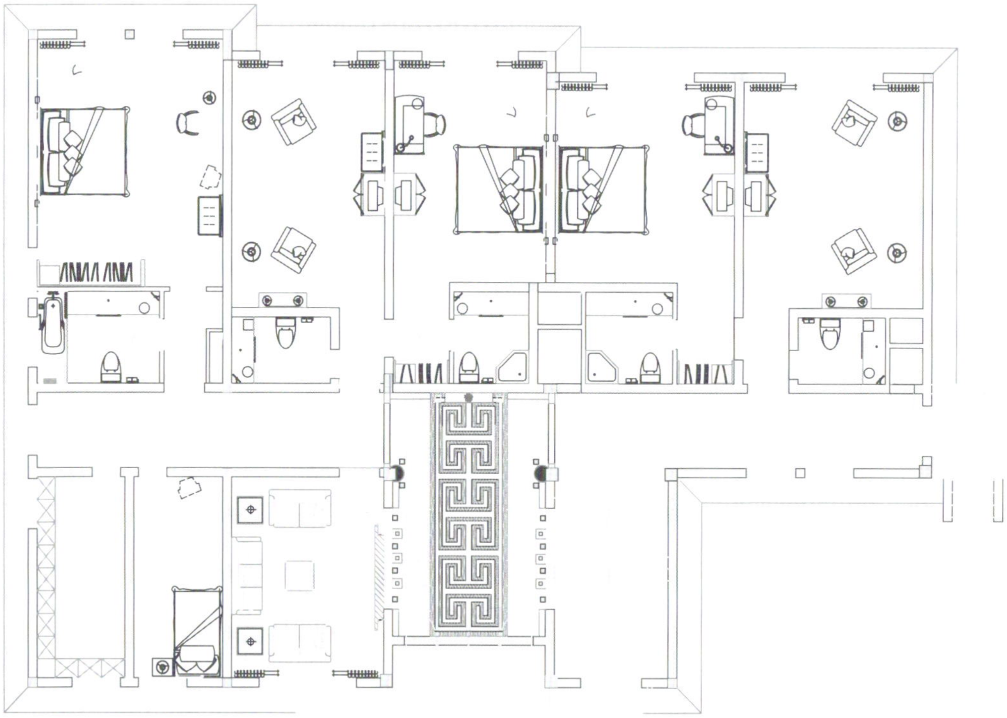

常熟国际饭店

CHANGSHU INTERNATIONAL HOTEL

设计负责人:马云飞　Architect: Ma Yunfei

设计参加人:顾跃强、王跃进　Collaborator: Gu Yueqiang, Wang Yuejin

建筑面积:5000m²　Building Size: 5000m²

工程造价:700万元　Construction Cost: RMB 7 million

获奖情况:荣获"最佳主题餐厅设计作品"奖　Awards: Best Theme Restaurant Design Prize

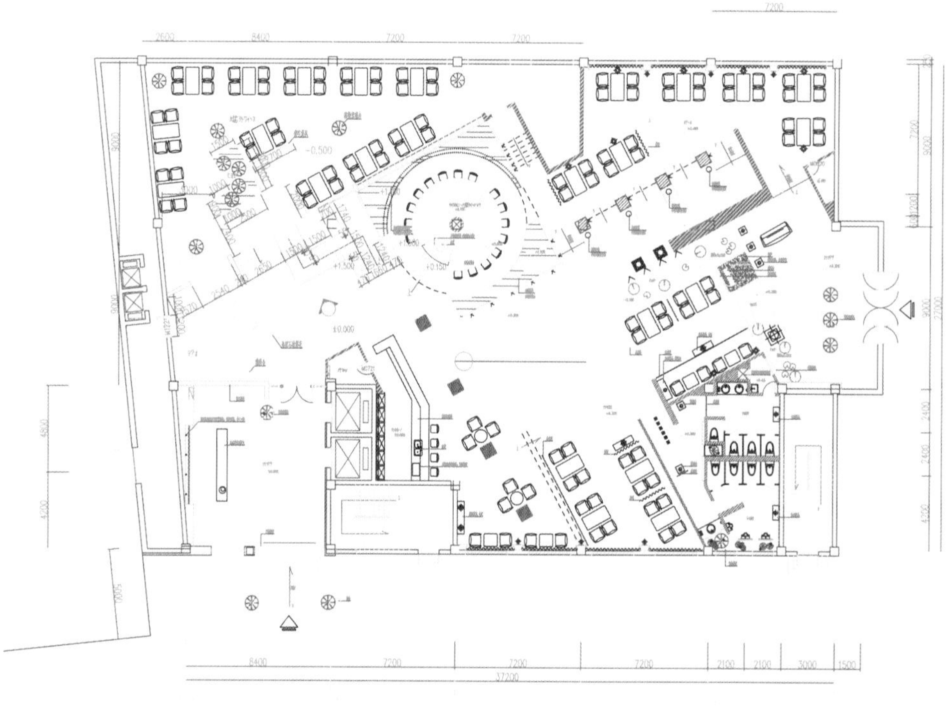

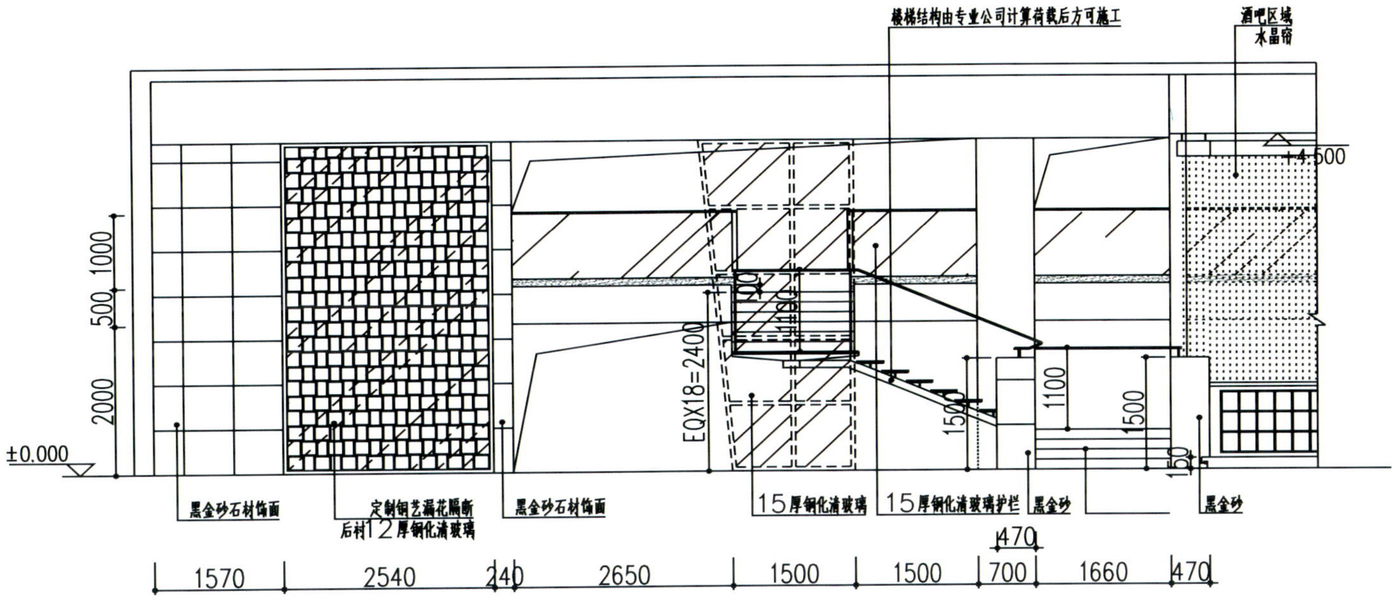
楼梯结构由专业公司计算荷载后方可施工
酒吧区域
水晶帘
+4.500
1000
500
2000
±0.000
EQX18=2400
1100
1500
150
150
黑金砂石材饰面
定制铜艺漏花隔断
后衬12厚钢化清玻璃
黑金砂石材饰面
15厚钢化清玻璃
15厚钢化清玻璃护栏
黑金砂
黑金砂
470
1570
2540
240
2650
1500
1500
700
1660
470

南昌东方花园

NANCHANG ORIENTAL GARDEN

设计负责人：周奕君　Architect: Zhou Yijun

设计参加人：李泉、雍忠孟、朱志霞　Collaborate: Li Quan, Yong Zhongmeng, Zhu Zhixia

建筑面积：46 500m²　Building Size: 46 500m²

工程造价：8000万元　Construction Cost: RMB 80 million

获奖情况：荣获“最佳酒店客房设计作品”奖　Awards: Best Hotel Guest Room Designing works Prize

对于一个利用办公楼改建的高星级酒店而言，除了合理分布酒店功能外，合适的空间尺度及比例、恰如其分的风格定位、彰显个性的视觉形象就显得至关重要。

南昌东方花园酒店的方案经过数轮探讨，最终形成了这样的定稿：把握都市高端客源的需求，以现代时尚的命题作为基础，打造一个豪华、精致但相对含蓄的现代化酒店。

在公共空间部分，将一些平实无奇的元素延伸成一个系列化的形象语言，反复穿插在各个空间里，组合出各呈异彩但又紧密联系的视觉关系。色彩也在一个基调内延伸，并点缀了少量的提神亮色，使整体气氛华丽细腻。

客房部分的房间形态复杂多变，在每个单独房型内仅把总体的形象语言再度延伸过来加以少量变化，气氛柔和雅致，简单中透出含蓄的精致。

For a star-rated hotel reconstructed from an office building, apart from a reasonable arrangement of the functions, it is of vital importance to have a proportional space size, a suitable style orientation and a visual image of distinguishing feature.

Through numerous rounds of discussion, the designing plan of Nanchang Oriental Garden Hotel came to the final version: to meet the need of the urban top guests, to create a luxurious, delicate and relatively reserved modern hotel on the basis of fashionable modern theme.

In the part of public space, the simple and insipid elements are developed into a series of image language, and repeatedly interweaved into each part of the space, forming respectively excellent but closely connected visual links. The colors also spread on the same keynote, with a few bright colors as the highlights, making the whole atmosphere magnificence and exquisite.

The guest rooms are of various styles. The overall image language is simply extended into each room with some minor changes, achieving a gentle, tasteful and simple atmosphere with reserved delicacy.

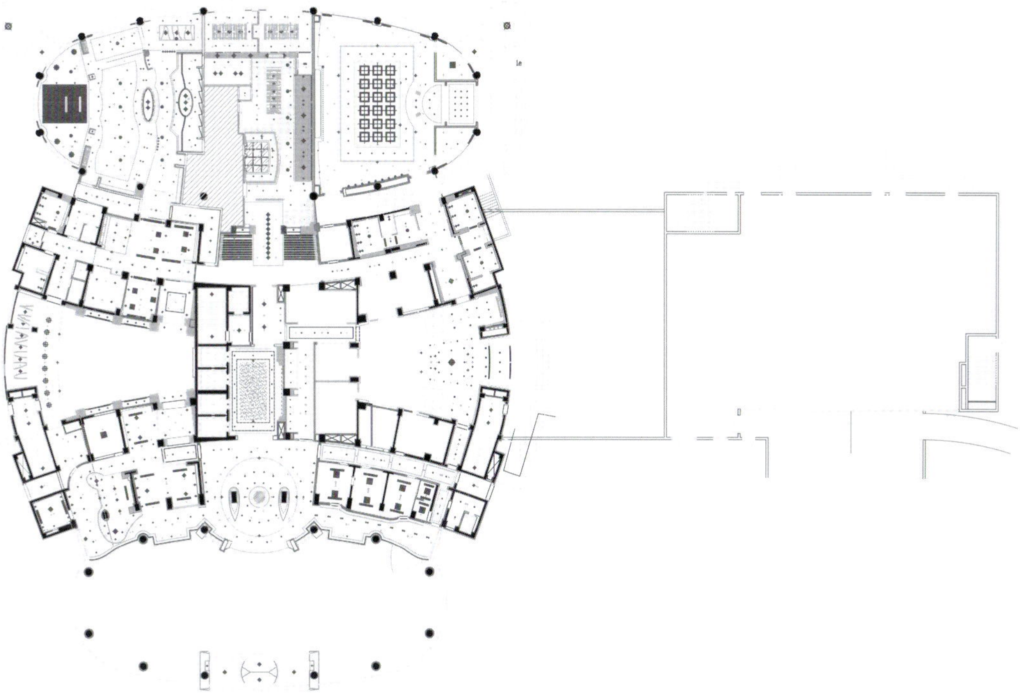

历史文化中心

HISTORICAL AND CULTURAL CENTER

设计负责人:陶峰、张东英、李伟 Architect: Tao Feng, Zhang Dongying, Li Wei
设计参加人:余长海、何剑、徐文渊 Collaborator: Yu Changhai, He Jian, Xu Wenyuan
建筑面积:30 000㎡ Building Size: 30 000m²

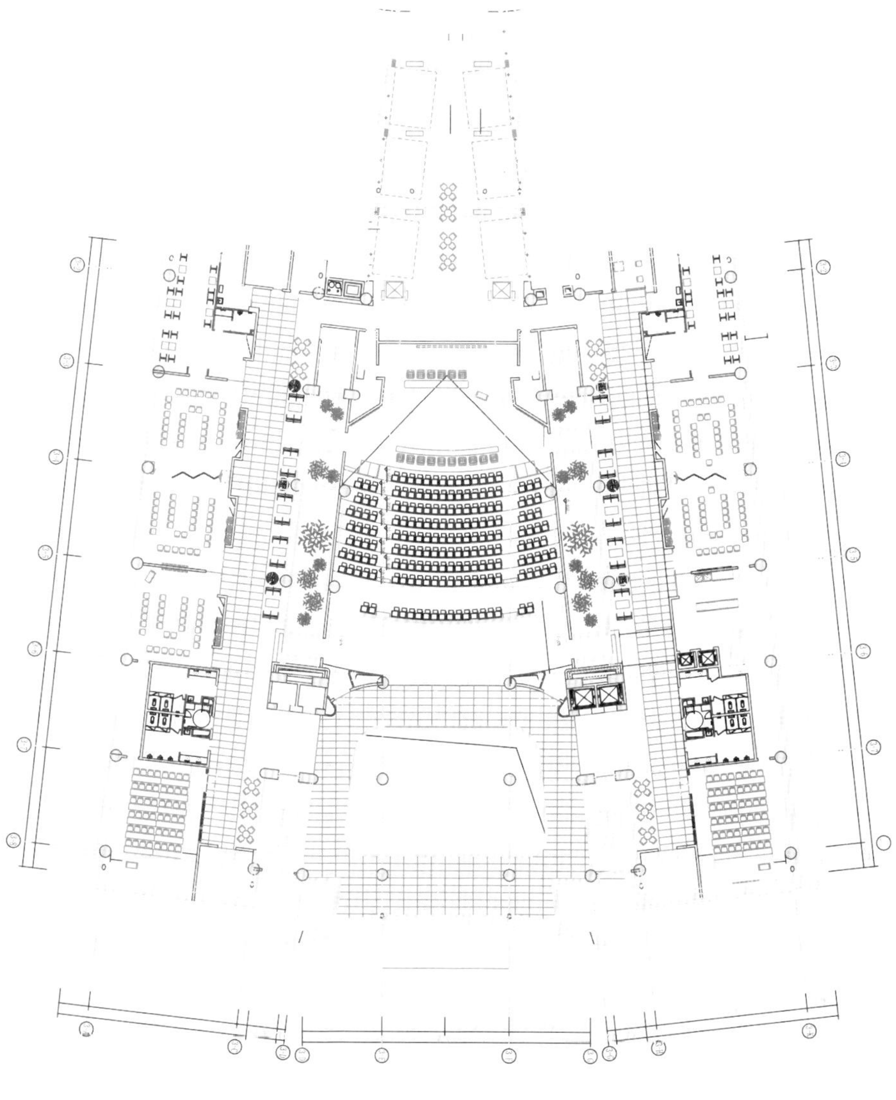

合肥中级人民法院办公楼

OFFICE BUILDING FOR INTERMEDIATE PEOPLE'S COURT, HEFEI

设计负责人:丁飞、陈刚 Architect: Ding Fei, Chen Gang

设计参加人:王辉、贝炜瑾、陆永花、鲁晴宇、张俊 Collaborator: Wang Hui, Bei Weijin, Lu Yonghua, Lu Qingyu, Zhang Jun

建筑面积 :约47 000㎡ Building Size: 47 000m^2

工程造价:4000万元 Construction Cost: RMB 40 million

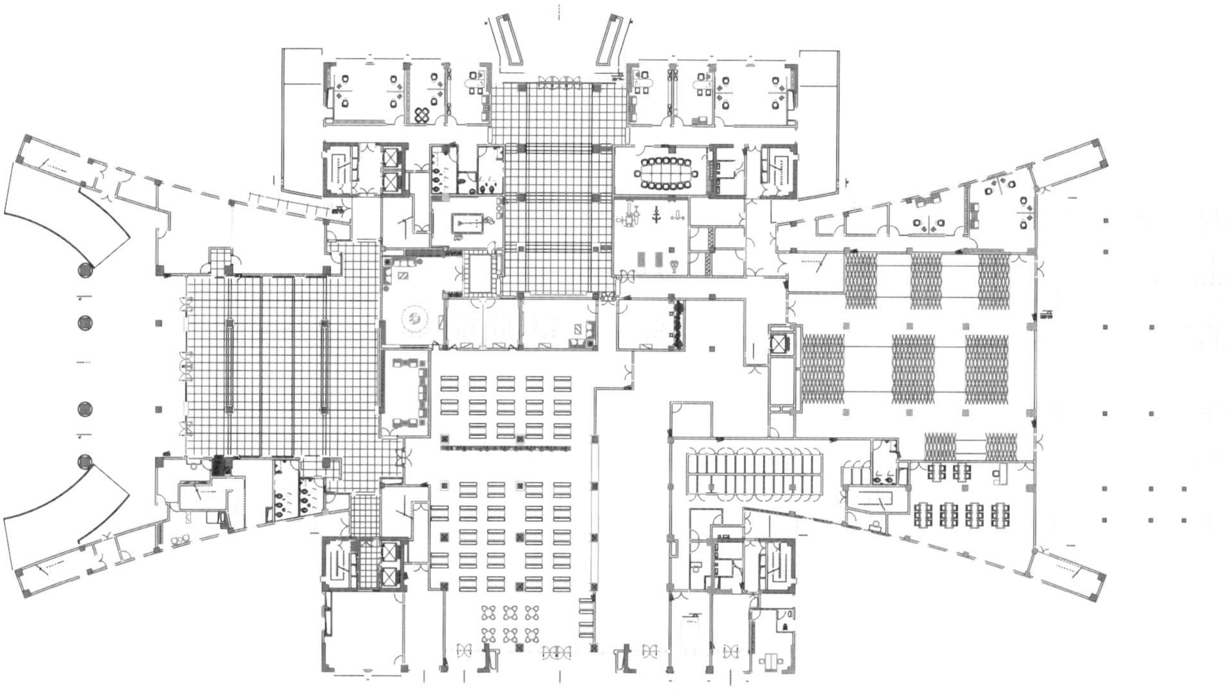

苏州国检大厦

OFFICE BUILDING FOR NATIONAL QUARANTINE BUREAU, SUZHOU

设计负责人：陈卫潭　Architect: Chen Weitan

设计参加人：盖晓君、史勤武、刘桦、王合胜　Collaborator: Gai Xiaojun, Shi Qinwu, Liu Hua, Wang Hesheng

建筑面积：35 000㎡　Building Size: 35 000m^2

工程造价：4100万元　Construction Cost: RMB 41 million

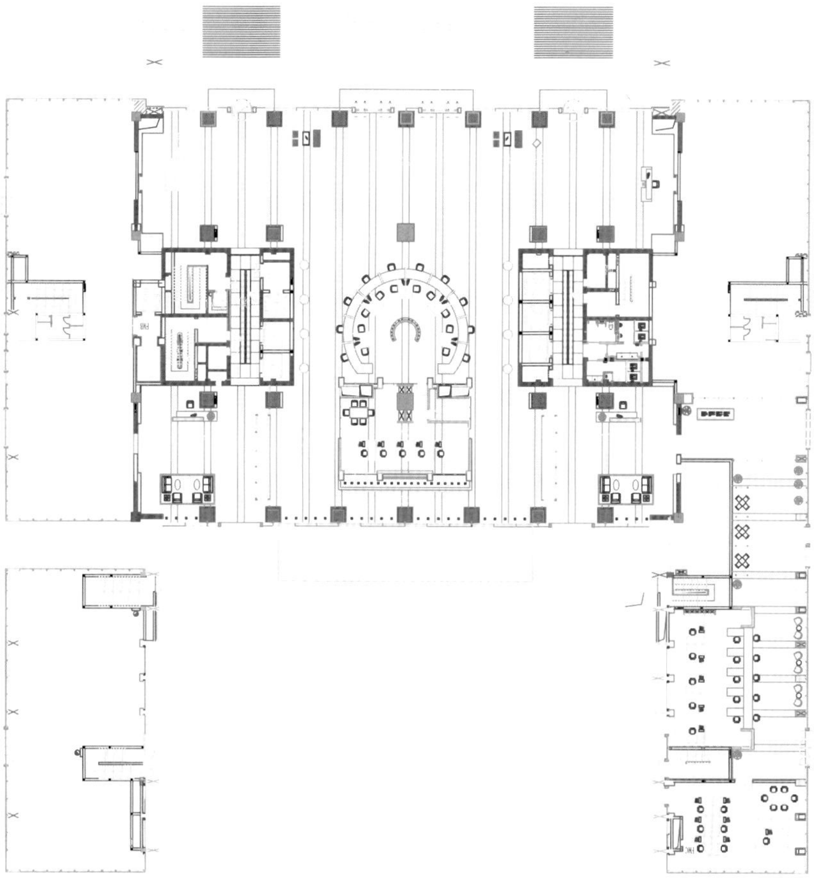

泰兴地税局办公楼

OFFICE BUILDING FOR TAIXING LOCAL TAXATION BUREAU

设计负责人：陈刚　Architect: Chen Gang

设计参加人：贝炜瑾、张燕、张颖、钱丽芬、张俊　Collaborator: Bei Weijin, Zhang Yan, Zhang Ying, Qian Lifen, Zhang Jun

建筑面积：12 000㎡　Building Size: 12 000m^2

工程造价：900万元　Construction Cost: RMB 9 million

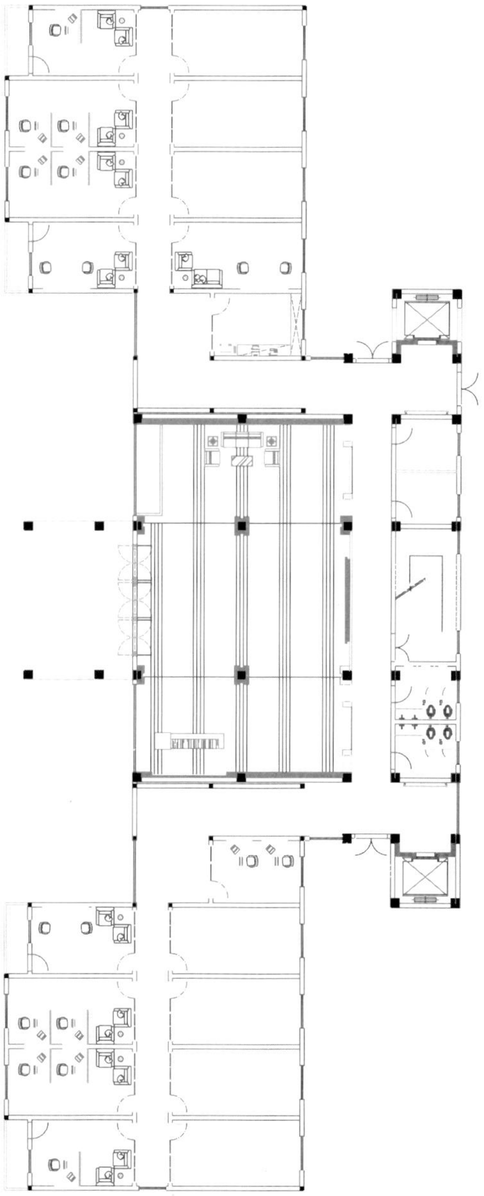

泰兴
地税

南京奥体中心科技楼

NANJING OLYMPIC SPORTS CENTER SCIENCE & TECHNOLOGY BUILDING

设计负责人：陈卫潭　Architect: Chen Weitan

设计参加人：史勤武、盖晓君、刘桦、王合胜　Collaborator: Shi Qinwu, Gai Xiaojun, Liu Hua, Wang Hesheng

工程造价：2600万元　Construction Cost: RMB 26 million

建筑面积：26 000㎡　Building Size: 26 000m^2

简介
Technology Center

浙江环球中心

ZHEJIANG GLOBAL CENTER

设计负责人:高超一、周斌 Architect: Gao Chaoyi, Zhou Bin

设计参加人:陈南、张建平 Collaborator: Chen Nan, Zhang Jianping

建筑面积:10 000㎡ Building Size: 10 000m^2

GLOBAL MALL
环球中心
ZHE JIANG GLOBAL MALL 浙江环球中
ZHE JIANG GLOBAL MALL 浙江环球中
GLOBAL MALL
问讯
INFORMATION

苏州园区路灯控制中心

SUZHOU INDUSTRIAL PARK STREET LAMPS SYSTEM CONTROL CENTER

设计负责人：李泉 Architect: Li Quan

设计参加人：徐杰、丁静、朱志霞、李娴、陈雪刚 Collaborator: Xu Jie, Ding Jing, Zhu Zhixia, Li Xian, Chen Xuegang

建筑面积：6000㎡ Building Size: 6000m^2

工程造价：1600万元 Construction Cost: RMB 16 million

苏州工业园区路灯控制中心坐落于苏州工业园区金鸡湖东侧，南施街西侧。总建筑面积约7900平方米。

整个设计在充分尊重原建筑及周边环境的基础上，利用色彩的对比，大的空间体块的穿插，巧妙地营造出简洁明快、现代时尚的空间氛围，体现出高效亲和的现代办公理念。在设计中通过细节的提炼和融合，塑造出了浓郁的人文气息。

The Street Lamps System Control Center is located on the east of Jinji Lake, and the west of Nanshi Street of Suzhou Industrial Park. Its total floor area is 7900m^2.

Fully respecting the original building style and surroundings, the design makes use of the color contrast, the interweave of big space block, and skillfully creates a simple, neat, pleasing and fashionable space atmosphere，embodying an efficient and friendly office idea. By refining and merging the details, the design is creating a rich human atmosphere.

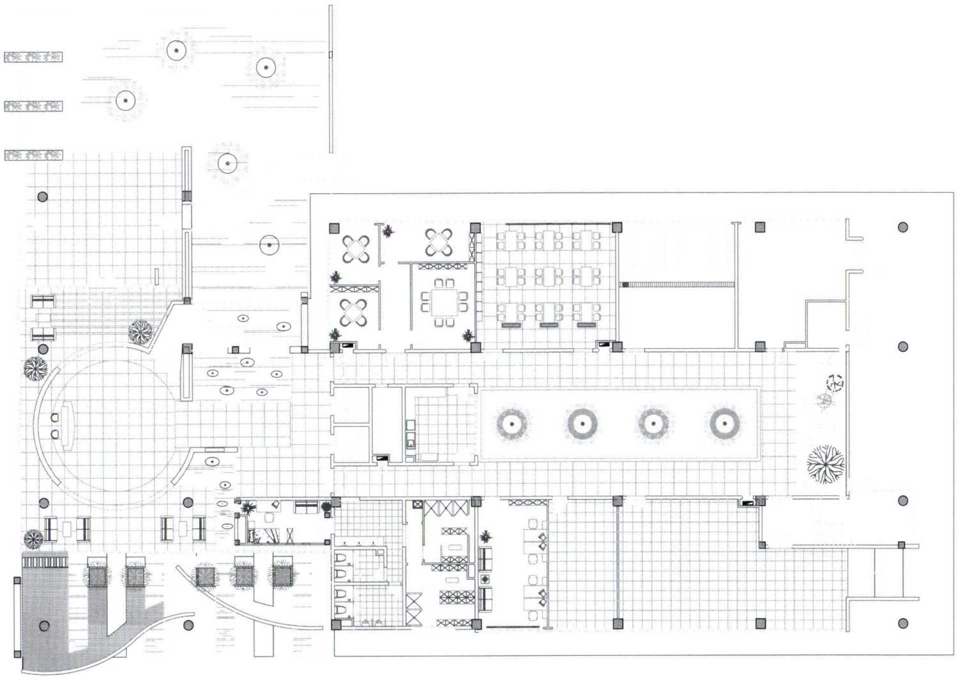

南京汉德森

NANJING HANDSON

设计负责人：李泉 Architect: Li Quan
设计参加人：徐杰、丁静、朱志霞、李娴、陈栋 Collaborator: Xu Jie, Ding Jing, Zhu Zhixia, Li Xian, Chen Dong
建筑面积：4000㎡ Building Size: 4000m^2
工程造价：800万元 Construction Cost: RMB 8 million

汉德森科技股份有限公司坐落于古城南京，它所在的南京江宁科技园，是集大学城、风景区和国家高新区为一体的综合开发区。

设计理念

科技+人文=有机高技术。
随着社会进步、经济与科技的飞速发展，纯粹的高科技美学已经满足不了人们的审美情趣，从而又慢慢开始关注地域文化、关注历史文脉、崇尚自然，两者结合，于是一种新的美学理论同时又是一种新的设计流派——“有机高技”诞生了，它使得技术更具有生命力，更为人性化。

意境与蕴涵

移步换景、借景、对景等江南园林设计手法与现代建筑的结合，一方面体现地方文化特色美，另一方面体现现代科技化建筑美，两者的有机结合充分体现了有机高技美学理论，使员工能专注工作，激发人的创造力。

Handson Science & Technology Co., Ltd. is located in Nanjing —— a city of great antiquity. Jiangning Science and Technology Park, Nanjing, where the company is nestled, is a comprehensive development zone, including university town, scenic spot and national hi-tech zone.

Design Philosophy

Technology + humanity = organic hi-tech
With the rapid development of society, economy and technology, the aesthetics of pure hi-tech can no longer meet the need of people's appreciation of beauty. Thereby, focusing on regional culture and historical context, together with conforming to nature, gradually becomes prevalent. By combining these two trends, organic hi-tech comes into existence—— a new aesthetic theory and a new school of architectural design, which makes technology more vigorous and more human.

Artistic Conception and Implication

The special beauty of regional culture and the beauty of the architecture with modern technology are both suggested by the combination of techniques of South Yangtze gardening, such as changing view at every turn, view borrowing, and opposite scenery etc, and the modern architecture. This combination which fully embodies the aesthetic theory of organic hi-tech, encourages the working staff to concentrate on their job and stimulates their creativity.

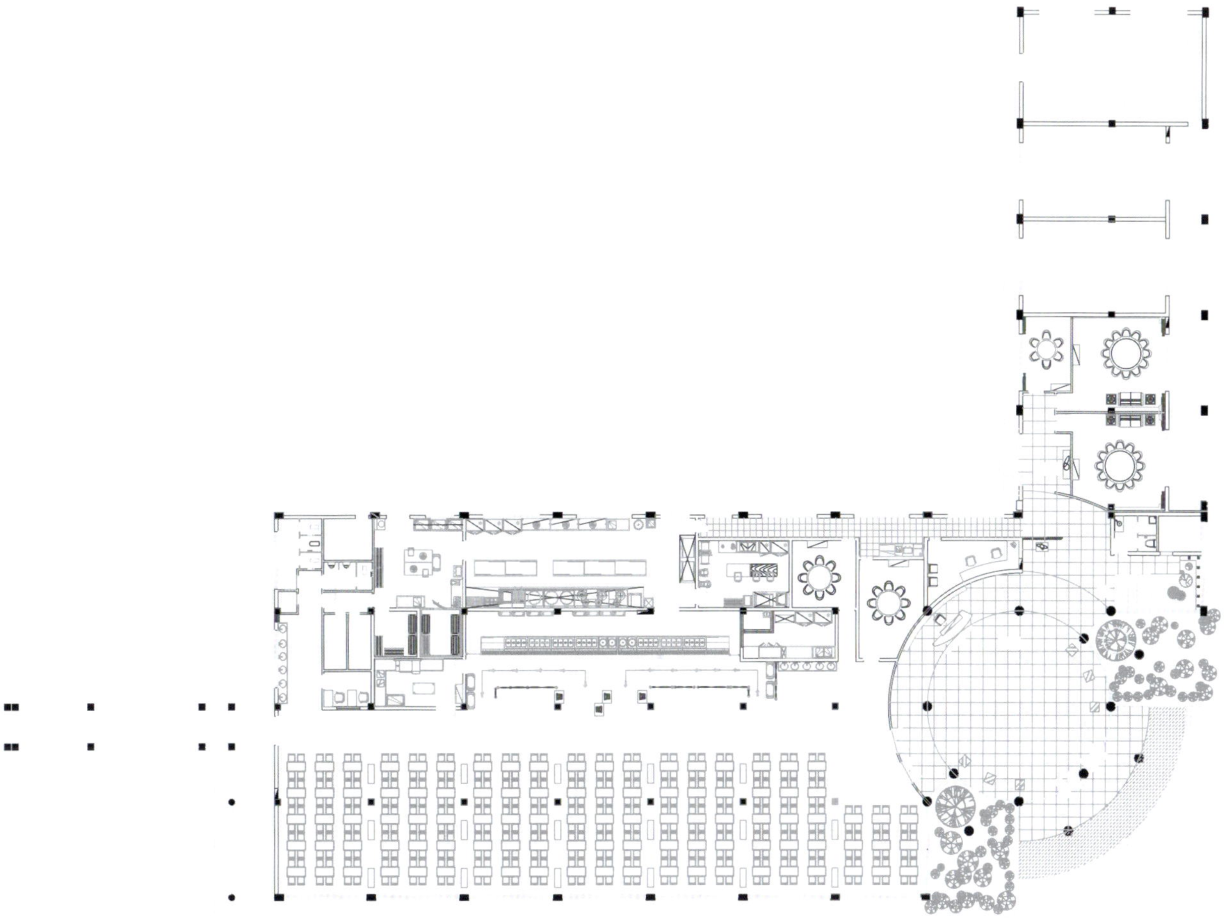

HANDSON SCIENCE

HANDSON SCIENCE

常熟虞山镇社区服务综合用房

COMPLEX BUILDING FOR YUSHAN TOWN COMMUNITY SERVICE CENTER, CHANGSHU

设计负责人:廖为军　Architect: Liao Weijun

设计参加人:包东群、李锋　Collaborator: Bao Dongqun, Li Feng

建筑面积:24 205㎡　Building Size: 24 205m^2

工程造价:2500万元　Construction Cost: RMB 25 million

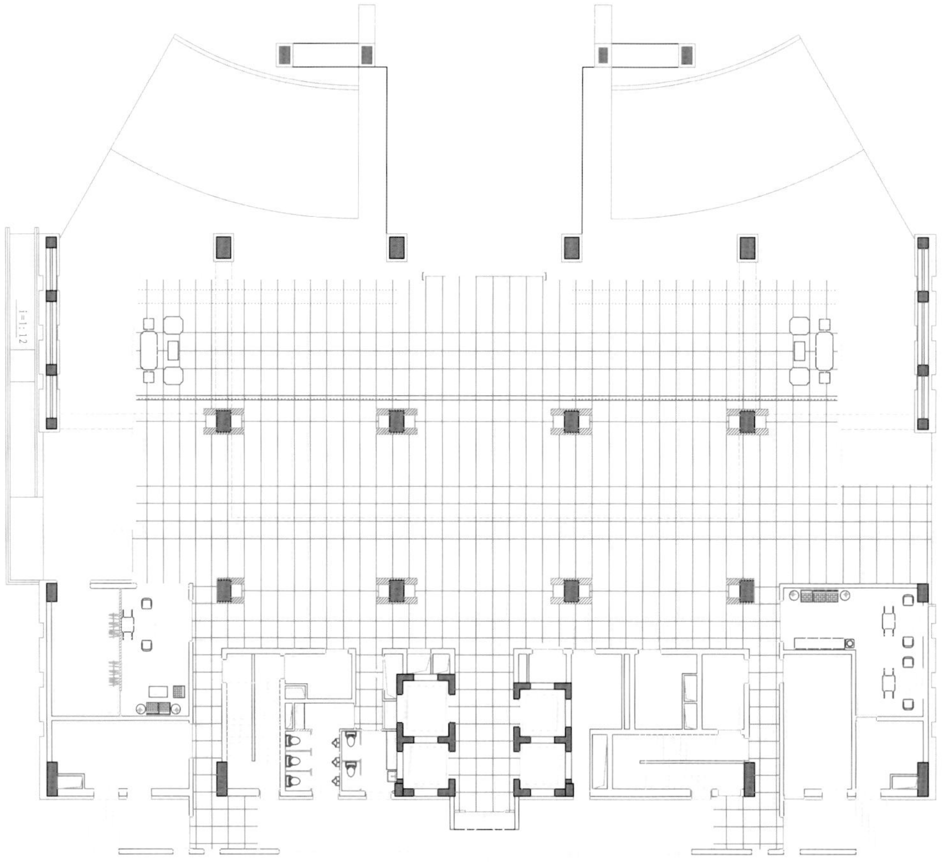

NETSCAPE
COMMUNICATOR

衢州财税大楼

QUZHOU FINANCE & TAXATION BUILDING

设计负责人: 程均清　Architect: Cheng Junqing

设计参加人: 周斌、张剑峰　Collaborator: Zhou Bin, Zhang Jianfeng

衢州市位于浙江西部，历史悠久，是国家级历史文化名城，四省通衢，资源丰富，生态优良。装饰设计中采用简洁、流畅、清新、典雅的设计风格，构成其办公文化。

本案在满足办公要求的同时，着重于个性与品位的表达，强调空间氛围的营造。在“办公氛围”方面，设计师认为，其本身并无固定的模式与格调，更多的是功能的体现，巧妙调动各不同功能区域形成对应互动，将具体的办公要求创意地、醒目地表达出来，以形成富有特色的办公氛围。

虚实对比强烈的设计手法，努力使整个方案轻秀不轻飘，严谨不严肃，庄重不凝重，力求符合国际建筑新趋势。

Quzhou city is located in the west of Zheing province. With its long history, Quzhou is a national-level historical and cultural city, which has convenient transport links with its four neighboring provinces, abundant natural resources and excellent ecological environment. The decoration design of Quzhou Finance & Taxation Building takes on a style of brevity, fluency, freshness and elegance, which constitutes the office culture.

Besides meeting the need of a working site, the design of the building also puts emphasis on expressing individuality, tastes and the building of unique space atmosphere. According to the designers, the office environment is with no fixed pattern or style in itself, it should largely expresses the functions. And the special office environment can be achieved through bringing different functional areas into play to form a corresponding interaction, and through creative and recognizable conveying of the specific office requirements.

This design of strong virtual-real comparison, is trying to make the whole building elegant rather than frivolous, precise rather than rigid, and stately rather than solemn, and aiming at conforming to the international new trend in architectural design.

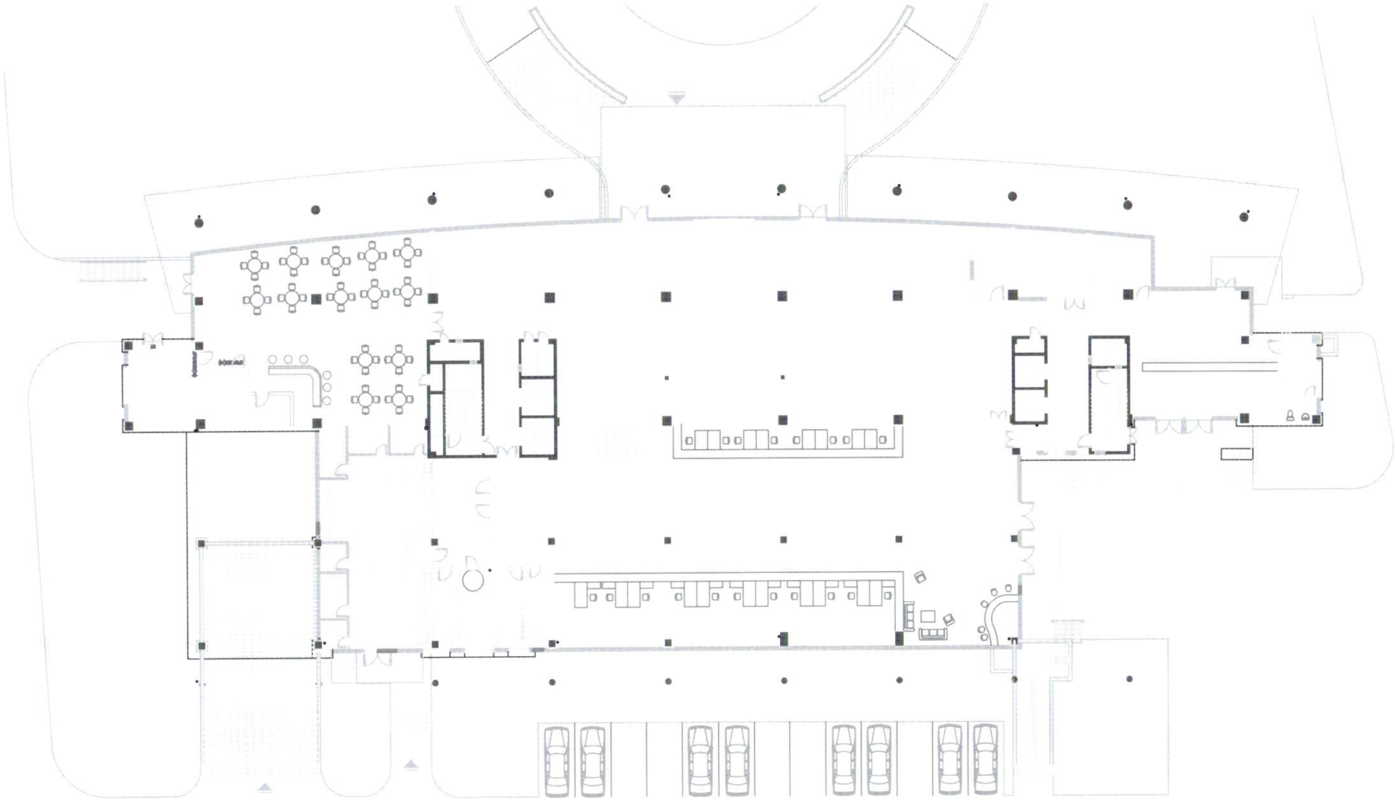

浙江科技馆

ZHEJIANG SCIENCE & TECHNOLOGY MUSEUM

设计负责人: 童向东　Architect: Tong Xiangdong

设计参加人: 程均清、张剑峰　Collaborator: Cheng Junqing, Zhang Jianfeng

建筑面积: 32 968m^2　Building Size: 32 968m^2

浙江科技馆设计中以“人”为主题来展开，并充分运用现代的装饰技术，烘托出现代科技馆的氛围，力求对不同空间在围绕“人”和“科技” 以及蓝色和黄色为主题的前提下创造出各个不同的让人有着震撼和耳目一新的感觉。

因原建筑设计中整个空间内大部分墙顶面多为玻璃幕墙，留给装饰的面并不多，所以我们只对入口墙面进行处理。左面墙面设置一块大型石材浮雕，金属板造型的售票亭上方设有一块电子屏幕，在大厅左侧休息区，设置了三个造型别致的电脑查询厅，内厅球体条形金属板造形并镶嵌条形发光体，下设科普展板，整个造型在体现“人”和“科技”的前提下，力求让参观者一进入馆内就能感受到震撼和耳目一新。吊顶镶嵌一条条发光体，给人一种向上和无限想象力的感觉。地面和柱面采用大面积白色岗石，整个设计风格简洁明快，富于现代感。

Developed on the theme of human, the design of Zhejiang Science & Technology Museum fully applies modern decoration techniques to set off the atmosphere of a modern Science & technology museum. Focusing on human and Science & technology and using the colors blue and yellow as the predominant colors, the design has brought about diverse, amazing and fresh atmosphere.

As most part of the wall top was originally glass curtain wall, leaving little space for further decoration, we only have the wall at the entrance decorated. The left wall is installed with a big stone relief sculpture. A digital information screen is put up on the top of a metal-bar booking office. In the rest area of the left lobby, we arranged three special computer-consulting rooms. The interior lobby is doom-shaped with metal bars and luminous strips inlaid. Some popular science exhibition boards are set up in the lobby. On the basis of embodying human and Science & technology, the whole design is to achieve an amazing and fresh style. With strips of luminosity inlaid onto the roof, it creates an initiative and imaginative feeling. With the ground and pillars covered with big white granites, the design looks simple, neat and pleasing, full of modern sense.

江苏工学院图书馆

LIBRARY OF JIANGSU POLYTECHNIC UNIVERSITY

设计负责人：包岳良　Architect: Bao Yueliang
设计参加人：黄强、陆雅萍　Collaborator: Huang Qiang, Lu Yaping
建筑面积:12 000m²　Building Size: 12 000㎡
工程造价：800万元　Construction Cost: RMB 8 million

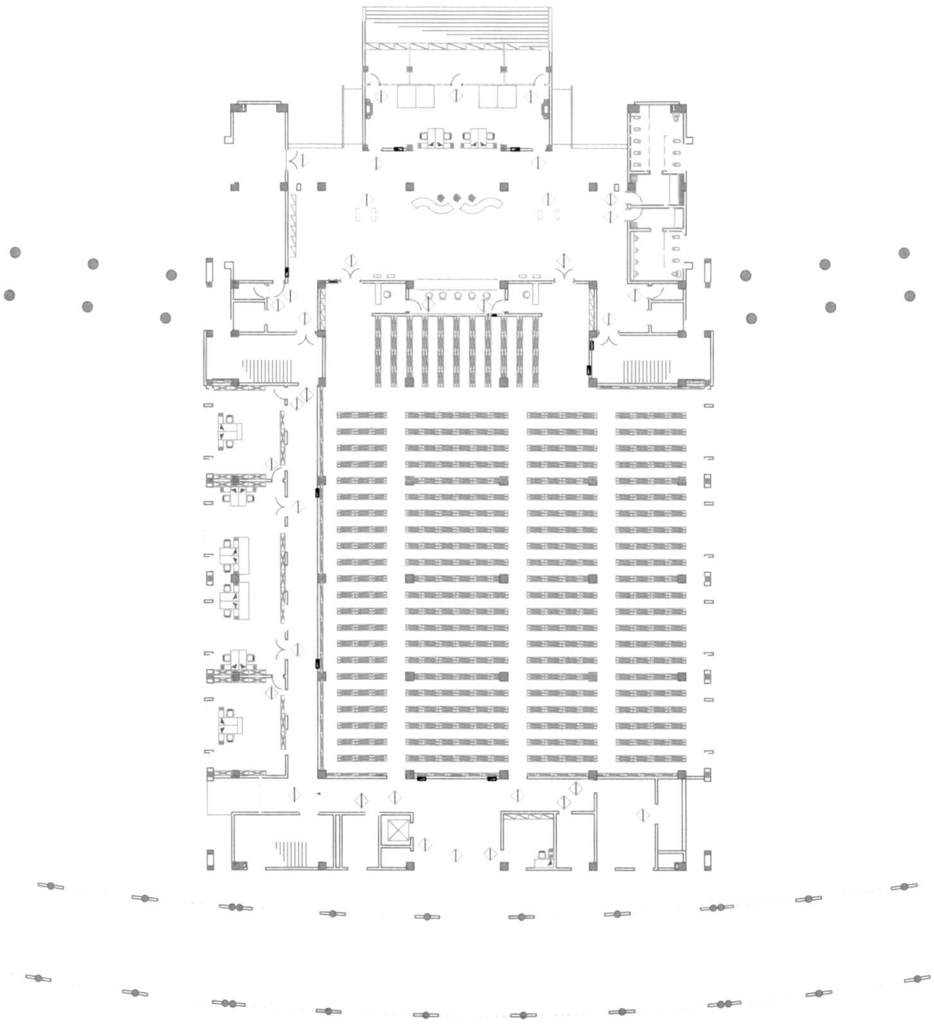

昆山龙腾光电企业办公楼

OFFICE BUILDING FOR KUNSHAN LONGTENG OPTICAL ELECTRICAL CO., LTD

设计负责人：丁飞、陈刚 Architect: Ding Fei, Chen Gang
设计参加人：王辉、陆永花、张颖、钱丽芬、郭海琦、许春来、张俊
Collaborator: Wang Hui, Lu Yonghua, Zhang Ying, Qian Lifen, Guo Haiqi, Xu Chunlai, Zhang Jun
建筑面积：约10 000㎡ Building Size: Around 10 000m^2
工程造价：1800万元 Construction Cost: RMB 18 million

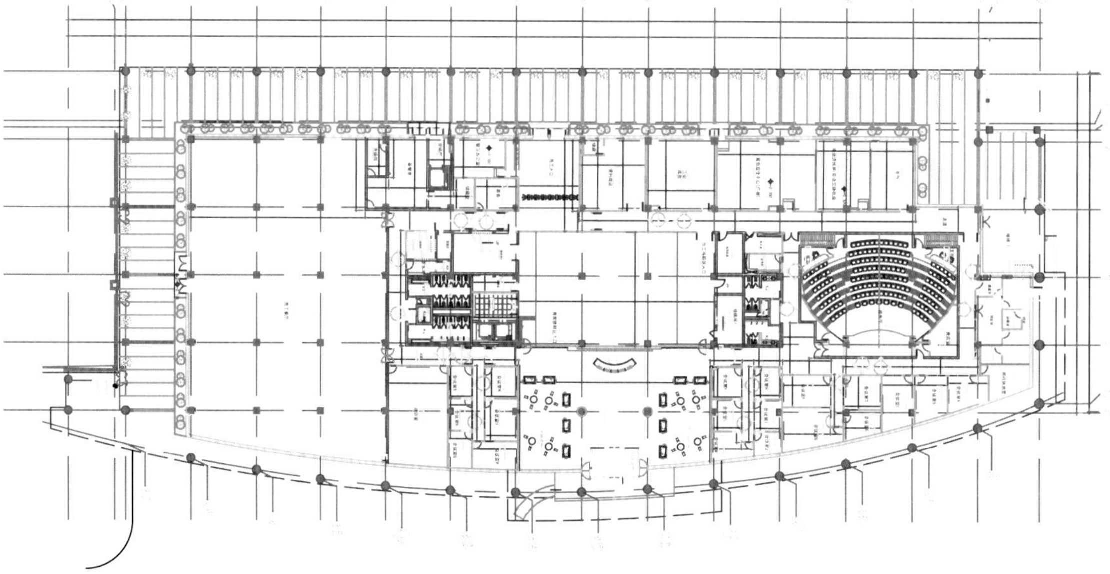

INFOVISION

常州市钟楼区人民法院

ZHONGLOU DISTRICT PEOPLE'S COURT, CHANGZHOU

设计负责人:包岳良 Architect: Bao Yueliang
设计参加人:赵创业、孙育 Collaborator: Zhao Chuangye, Sun Yu
建筑面积:14 000㎡ Building Size: 14 000m²
工程造价:1000万元 Construction Cost: RMB 10 million

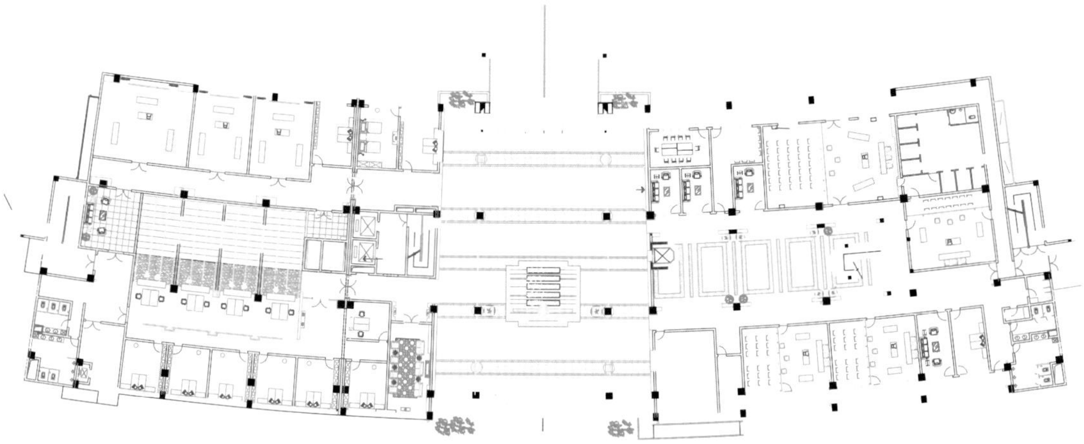

业务受理
业务受理

立案窗口（2）
案窗口（3）

广东发展银行杭州分行

GUANGDONG DEVELOPMENT BANK, HANGZHOU BRANCH

设计负责人：童向东　Architect: Tong Xiangdong
设计参加人：程均清、张剑峰　Collaborator: Cheng Junqing, Zhang Jianfeng
建筑面积：24 615m² Building Size: 24 615m²

办公门厅，设计上力求创造一个自然的、文化的、简洁大方、庄重美观的环境。用材上采用木纹石材与木材相结合，墙上石材造型的银行标识，增加了门厅的艺术氛围，同时又点明了大楼的性质。

电梯厅既是人员进出的主要地方，又是垂直交通的枢纽地带，在人们的视觉中占有相当的比例。因此，从电梯门套、地面的拼花、墙面的装饰、吊顶的造型及细小的微妙之处无不体现着设计者的匠心独运。带有银行标志的烤铀玻璃墙面以及和建筑外立面相呼应的发光墙面与顶面接合的造型，既增加了电梯厅的空间感，又象征了广发银行的发展将更加辉煌。

17层中厅作为银行高层的办公楼层，深色平静的水池，白色的大理石，抽象的雕塑，周围简洁的木纹石墙地面以及钢结构玻璃楼梯和一直向上的木栅造型，无不体现出银行高层的品质和广发银行的精神。带给人一种充满艺术和想象的灵动空间。

黄金俱乐部作为银行高层开会和接待的场所，在该层设置了一个封闭的20人左右的会议室以及洽谈、接待、休闲的开敞场所。会议室圆形的通光云石吊灯和圆形的会议桌，以体现广发银行高层的公开、平等、团结的工作作风，墙面艺术玻璃和木条造型以体现一种平和的气氛。整个俱乐部空间力求创造一个尊贵、祥和又带有银行特色的活动空间。

The design of entrance hall is trying to create a natural and elegant environment — cultural, simple, graceful and stately . As for the materials, timber stone and wood are adopted. The stone symbol of the bank on the wall adds to its artistic atmosphere of the hall and indicates its nature.

Lift lobby is not only an important entrance-exit area, but also the hub of vertical communication. It is playing an important part in people's eyes. So the lift door frame, the floor mosaic, the wall decorations, the modeling of the suspended ceiling and the other trivial parts all reflect the designer's unique designing style. The glazed glass wall with the bank symbol on and the joint modeling of the luminous wall and ceiling that matches with the exterior wall not only add to the space sense of the lift lobby, but also symbolize the prosperous future of the bank.

The central hall on the 17th floor is the office center for the senior officials of the bank. The peaceful dark water pool, the white marble, the abstract sculpture, the clear timber stone wall and ground, the steel-glass stairs and the upward wooden fence modeling, all embody the senior officials' quality and the bank's spirit, bringing about an artistic and imaginative atmosphere.

Golden Club is the conference and reception center for the senior officials of the bank. It contains a closed conference room with the capacity of 20 people and an open area for chat, reception and relaxation. The round translucent marble pendant lamp and the round conference table represent the senior officials' working style: openness, equality and unity. The artistic glass and wooden strip modeling of the wall therefore stands for a peaceful atmosphere. The club's design is creating a noble and serene working space with the bank's characteristics.

广东发展
GUANGDONG

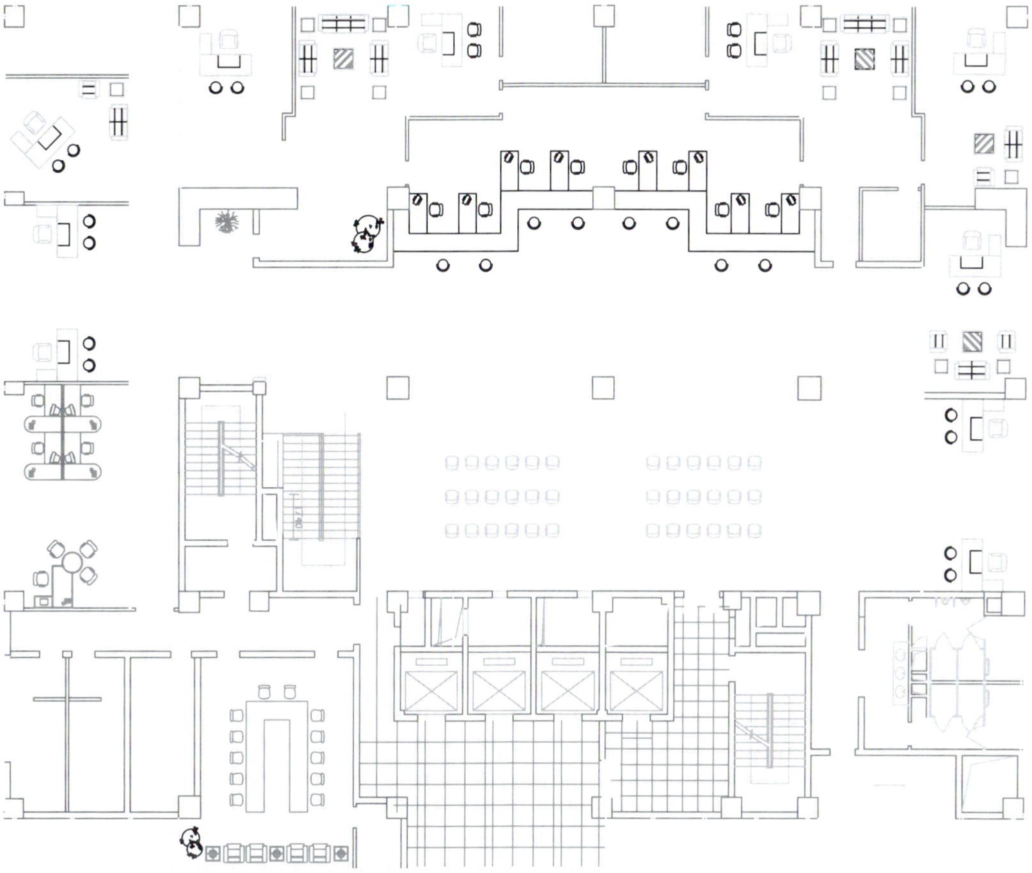

南通金太阳床上用品有限公司综合大楼

COMPLEX BUILDING FOR NANTONG GOLD SUN BEDCLOTHES CO., LTD

设计负责人：黄健 Architect: Huang Jian

设计参加人：汤宇峰、王智弘、蒋维 Collaborator: Tang Yufeng, Wang Zhihong, Jiang Wei

建筑面积：8000m² Building Size: 8000m²

工程造价：700万元 Construction Cost: RMB 7 million

获奖情况：江苏省“柏森杯”一等奖 Awards: Bai Sen Cup First Prize, Jiangsu province

南通金太阳床上用品有限公司是当地一家颇具实力的企业。大楼分为展示和办公两大功能，在设计中根据业主的要求和企业的特点，提出以下设计理念：

大楼要体现企业形象。在设计中运用水、圆等元素体现金太阳家纺行业的文化特征；展示区精心分区，有效地把产品种类予以明确，单元展区营造家庭卧室气氛；在造型上强调简洁明快，运用面的交错和穿插丰富空间；运用现代材料和布料的有机结合体现出时尚。

Nantong Gold Sun Bedclothes Co., Ltd is a local famous company. The building is for exhibition and office purposes. In the light of proprietor' s demand and the company' s characteristics, the design puts forward the following ideas.

The design should embody the company' s image. So elements like water and circles are used to represent the corporate culture of Gold Sun Bedclothes Co., Ltd. The exhibition area is elaborately demarcated and effectively made clear of the types of the products. And each exhibition unit is designed to create an atmosphere of a bedroom in a family. The design is pursuing a simple, neat and pleasing style. The crisscross and interweave of planes are used to enrich the space while the organic combination of modern materials and clothes exhibits the fashion.

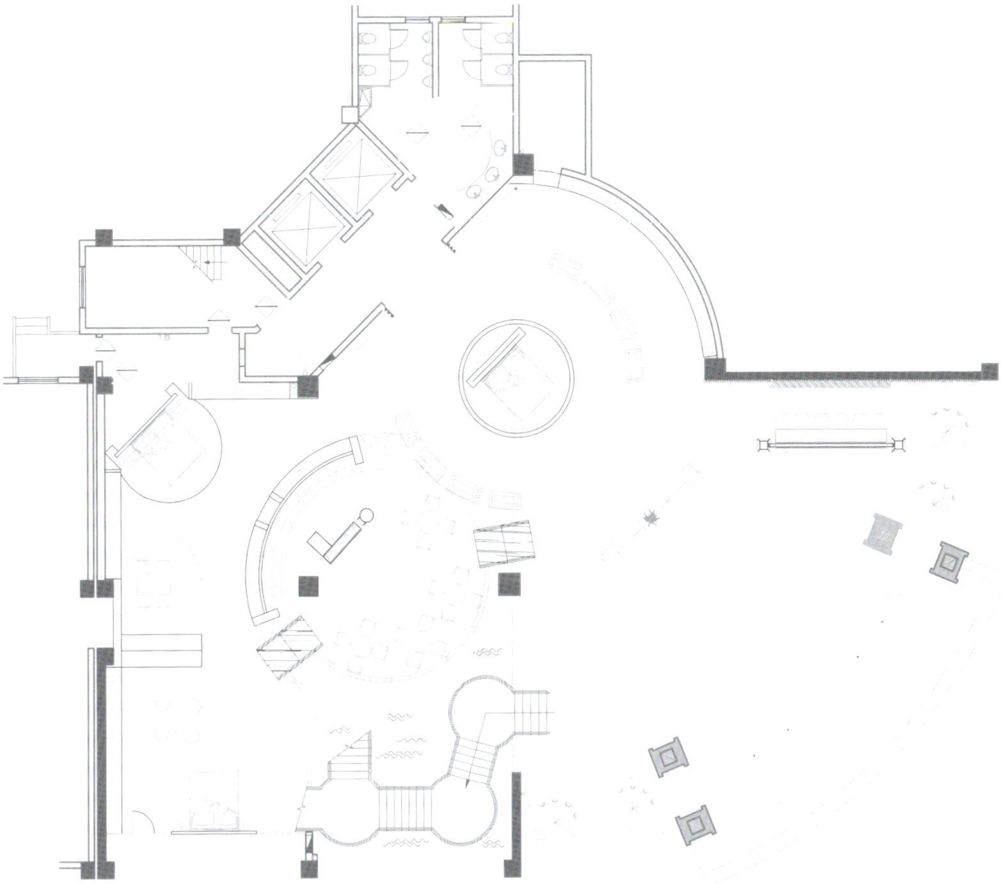

GOLDSUN

GOLD SUN
GOLDSUN
GOLDSUN

江苏紫罗兰家纺综合大楼

COMPLEX BUILDING FOR JIANGSU VIOLET HOME TEXTILE CO., LTD

设计负责人:黄健　Architect: Huangjian

设计参加人:王智弘、蒋维　Collaborator: Wang Zhihong, Jiang Wei

建筑面积:12 000m^2　Building Size:12 000m^2

工程造价:1000万元　Construction Cost: RMB 10 million

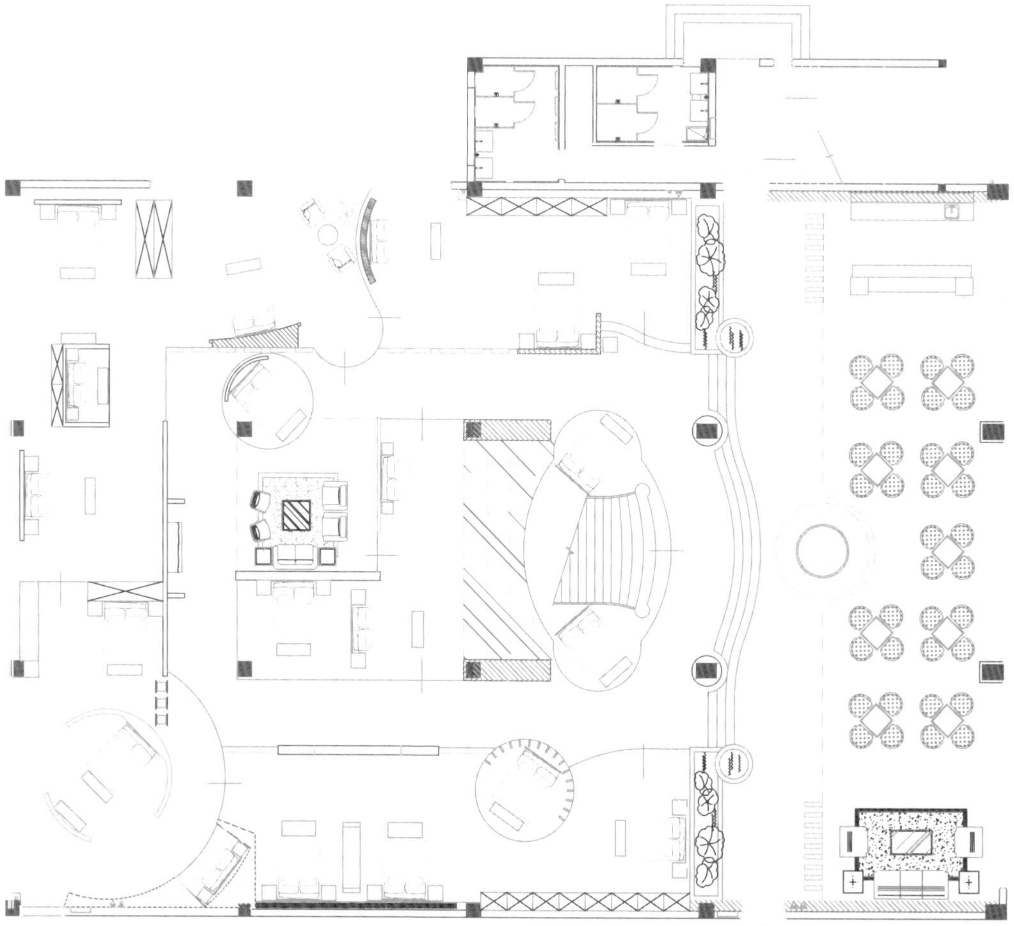

长兴县行政综合楼

COMPREHENSIVE ADMINISTRATION BUILDING, CHANGXING COUNTY

设计负责人：程均清　Architect: Cheng Junqing

设计参加人：童向东、程均清　Collaborator: Tong Xiangdong Cheng Junqing

建筑面积：30 000m^2　Building Size: 30 000m^2

一层公共空间作为大楼对外接待的一个主要窗口。为了将室内设计与建筑设计融为一体，在设计中采用了大量的石材、玻璃及铝板等现代的建筑材料，用简洁、明快的设计手法，以求在形上与外立面统一，建筑室内色调基本以黑、白、灰等冷色调为主，并继续沿用理性现代词汇表现建筑简洁明快的特点。在大厅屋顶的中央开口，使更多的日光从顶部进入室内空间，消除了由屋顶平面造成的沉重和压迫感，设计师的两个方案均采用穿孔铝板飘板造型，在不影响室内采光的前提下，丰富采光顶造型，在强日光照射时起到光的调节作用。夜间在下方打光，产生漫反射，使大厅的光线柔和。立面处理方面采用了磨砂玻璃与不锈钢线条相结合的构成手法，从视觉上消除三层墙体过于沉重的体量。在门厅正面采用浮雕，与玻璃的造型形成强烈的虚实对比。地面拼花采用大块面与线条的穿插将大厅贯通在一起，并显出强烈的空间导向性，打造出一个现代大气的公共空间。

The public space on the first floor is the main reception area of the building. In order to make the interior design and the construction design in harmony, a lot of modern building materials like stone, glass and aluminum board are widely used in a simple, neat and pleasing way, to pursue the unity with the outside wall. The interior design is mainly based on the cold colors like black, white and grey, and continues to use rational modern architectural words to embody the simple, lively features of the building. The design leaves the central part of the entrance hall roof open, which can let in more light and get rid of the depressing sense owing to the big-sized plane ceiling. The two designing plans both adopt punch aluminum slab structures to beautify the daylight modeling without affecting the natural daylight in, and to modulate the light when it is too dazzling. In the evening the lights will shine onto the roof from below, producing diffuse reflection, making the light in the hall soft and gentle. In terms of the vertical plane, the skill of frosted glass and stainless strips is applied to remove visually the heavy dimension of 3-layer walls. Facing the entrance, the relief sculpture and the glass modeling form a sharp real-fictitious contrast. The ground mosaic is finished with the interweaving of big squares and lines to run through the whole hall, showing clearly the guiding effect, and creating a modern and splendid public space.

7

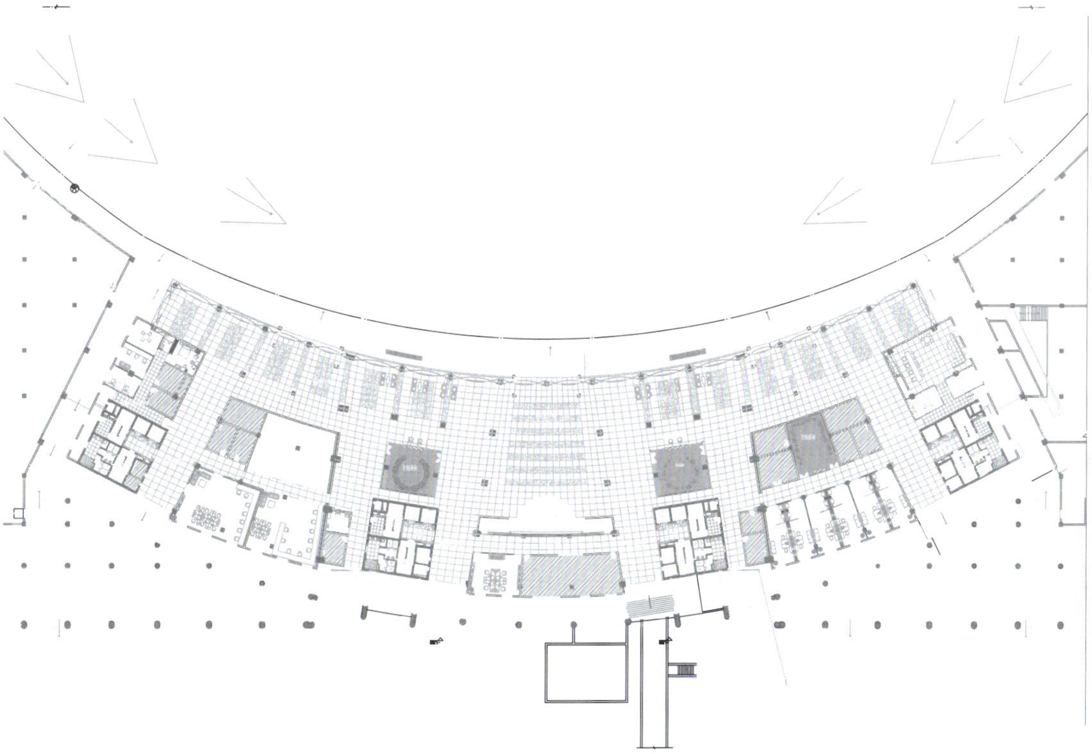

滨江国际大厦

BINJIANG INTERNATIONAL BUILDING

设计负责人：廖为军　Architect: Liao Weijun

设计参加人：包东群、李锋　Collaborator: Bao Dongqun, Li Feng

建筑面积：25 000m² Building Size: 25 000m²

工程造价：3000万元　Construction Cost: RMB 30 million

获奖情况：江苏省"建伟杯"建筑装饰优秀设计（方案）竞赛优秀奖

Awards: Recognition Award in Jian Wei Cup Contest of Excellent Design for Architectural Decoration, Jiangsu province

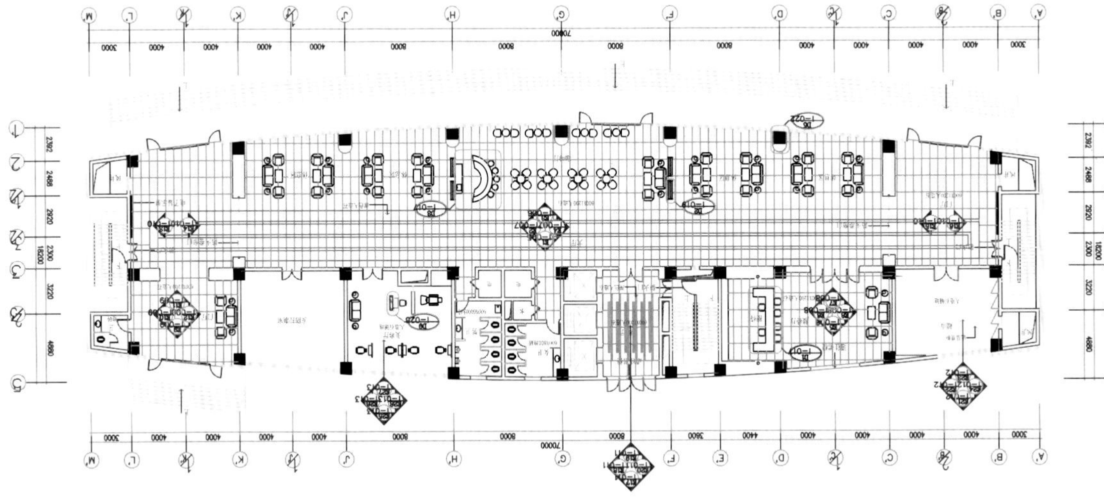

CHANGSHU

海安信用联社大厦

HAIAN CREDIT COOPERATIVES UNION BUILDING

设计负责人:段秀丽　Architect: Duan Xiuli

设计参加人:徐敏、唐一丰、陆红娟、颜永春　Collaborator: Xu Min, Tang Yifeng, Lu Hongjuan, Yan Yongchun

建筑面积:12 000m²　Building Size: 12 000m²

工程造价:890万元　Construction Cost: RMB 8.9 million

获奖情况:江苏省“柏森杯”三等奖　Awards: Bai Sen Cup Third Prize, Jiangsu province

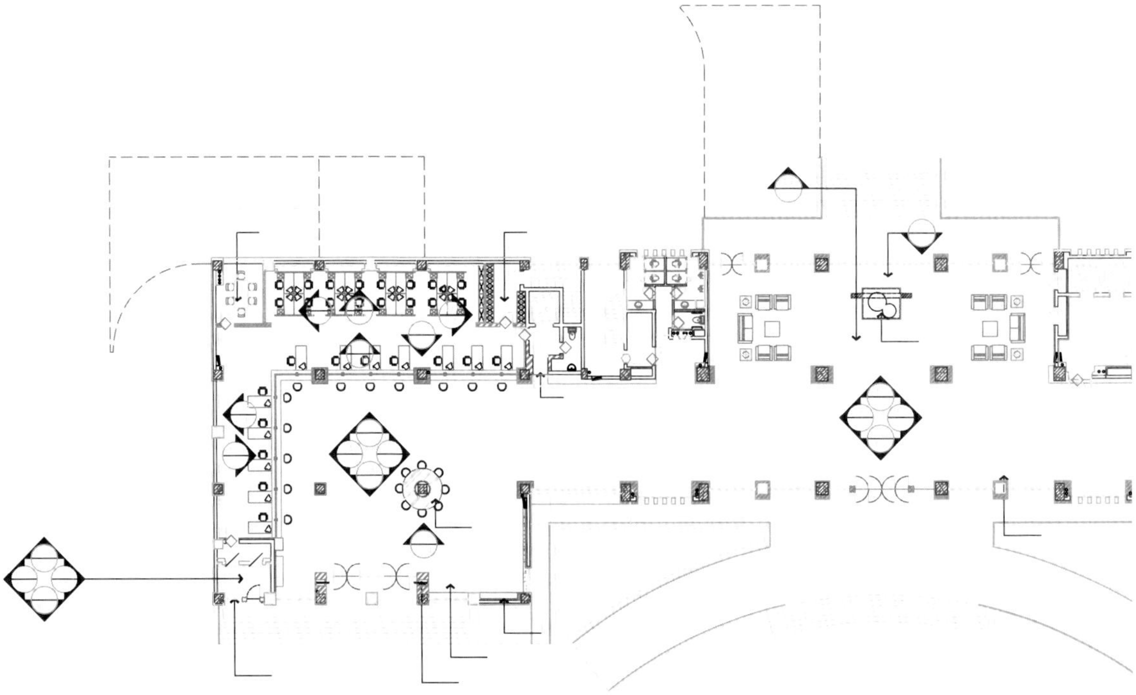

12345678910
12345678910

城管局
质监局
国税局
国税局

LCD
MULTIMEDIA PROJECTOR

苏州吴中区法院

SUZHOU WUZHONG DISTRICT PEOPLE'S COURT

设计负责人:段秀丽　Architect: Duan Xiuli
设计参加人:陆红娟　Collaborator: Lu Hongjuan
建筑面积:12 000m^2　Building Size: 12 000m^2
工程造价:960万元　Construction Cost: RMB 9.6 million

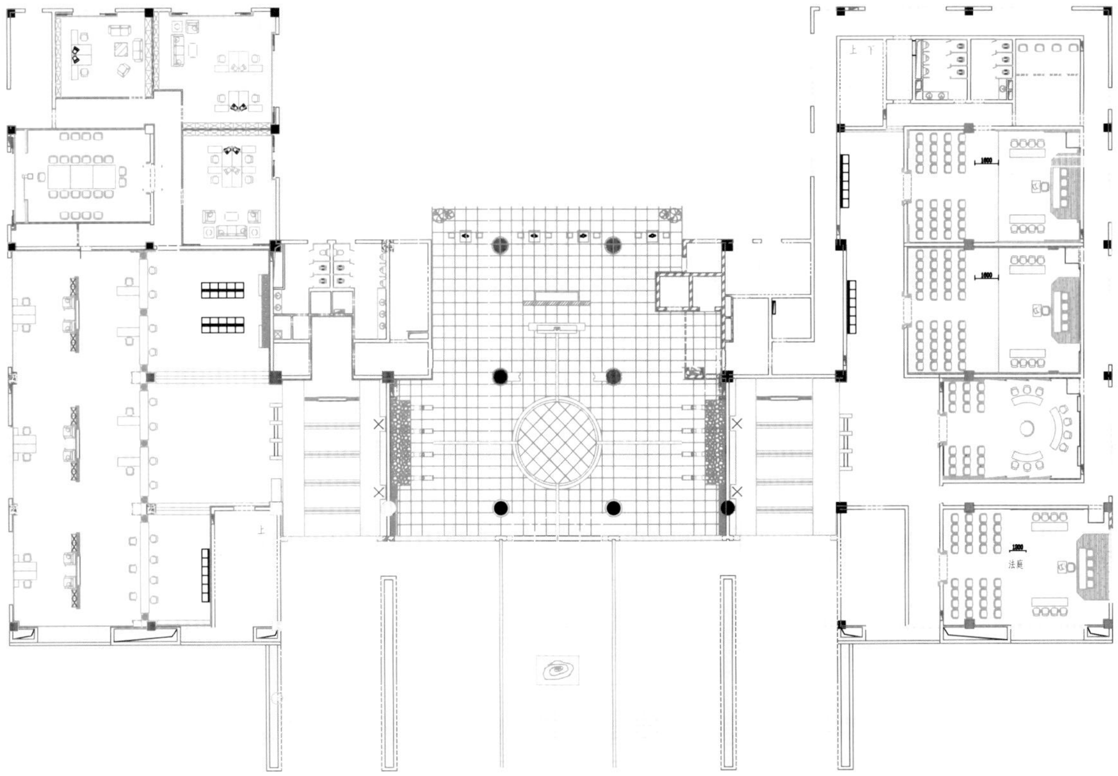

案件受理窗口 2

艾力枫高尔夫俱乐部

AILIFENG GOLF CLUB

设计负责人：谢天　Architect: Xie Tian
设计参加人：鲁涛、黄俊、张剑峰　Collaborator: Lu Tao, Huang Jun, Zhang Jianfeng
建筑面积：15 000㎡　Building Size: 15 000m^2

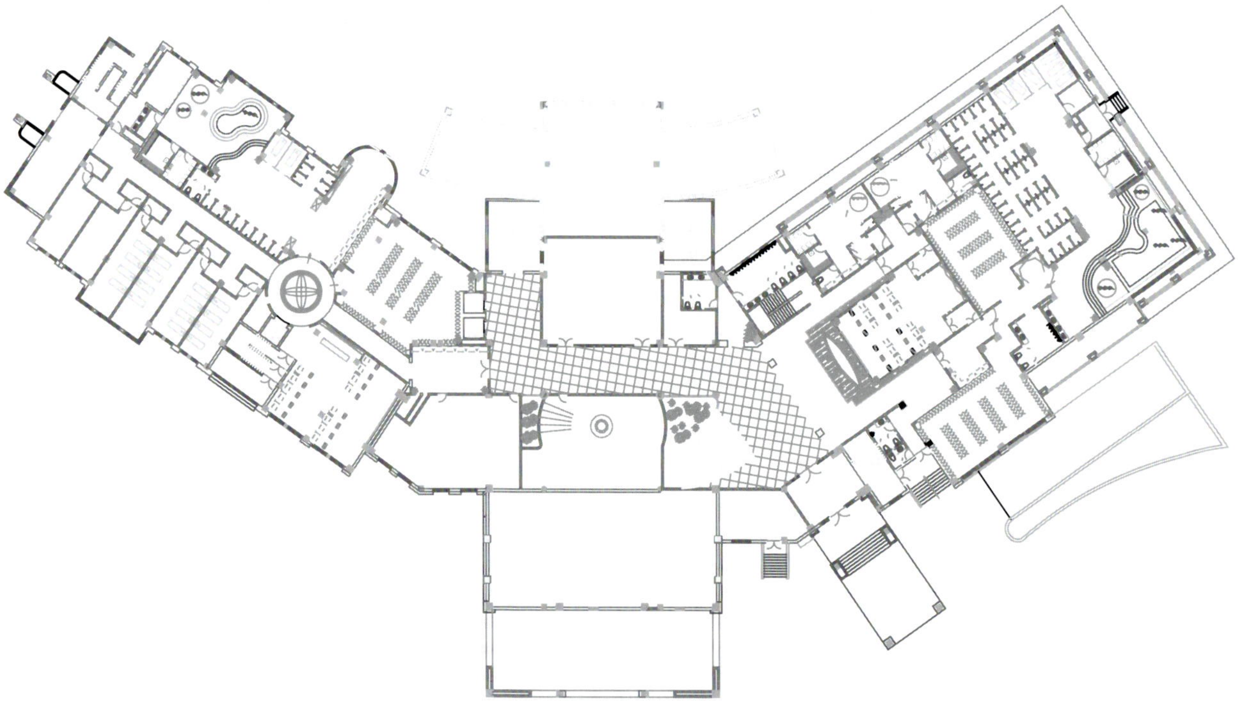

尼盛棕榈湾会所

NISHENG PALMS BAY CLUB

设计负责人: 潘芳　Architect: Pan Fang

设计参加人: 张艳娜、陈洁　Collaborator: Zhang Yanna, Chen Jie

建筑面积: 1000㎡　Building Size: 1000m²

工程造价: 200万元　Construction Cost: RMB 2 million

获奖情况: 江苏省“柏森杯”三等奖　Awards: Bai Sen Cup Third Prize, Jiangsu province

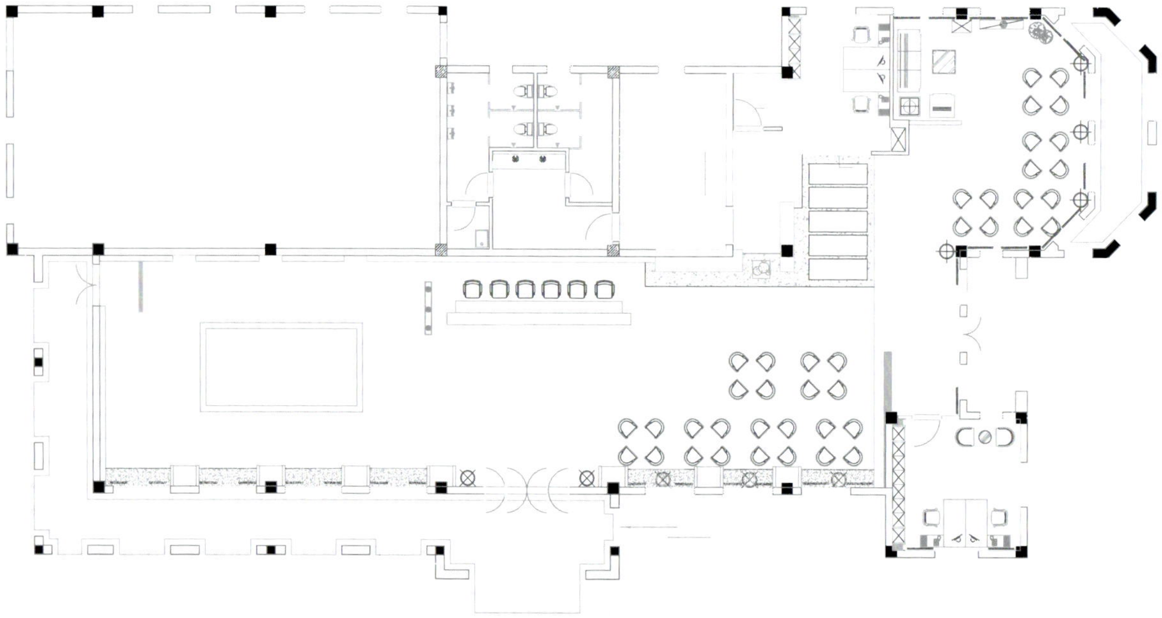

华新国际七期样板房

HUAXIN INTERNATIONAL (PHASE Ⅶ) MODEL ROOMS

设计负责人：段秀丽　Architect: Duan Xiuli

设计参加人：徐敏、颜永春　Collaborator: Xu Min, Yan Yongchun

获奖情况：江苏省“柏森杯”三等奖　Awards: Bai Sen Cup Third Prize, Jiangsu province

兴化天宝花园会所

XINHUA TIANBAOHUAYUAN CLUB

设计负责人：丁飞、陈刚
设计负责人：潘芳　Architect: Pan Fang
设计参加人：张艳娜，曹海勇　Collaborator: Zhang Yanna, Cao Haiyong
建筑面积：$5000m^2$　Building Size: $5000m^2$
工程造价：10 000 000万元　Construction Cost: RMB 10million

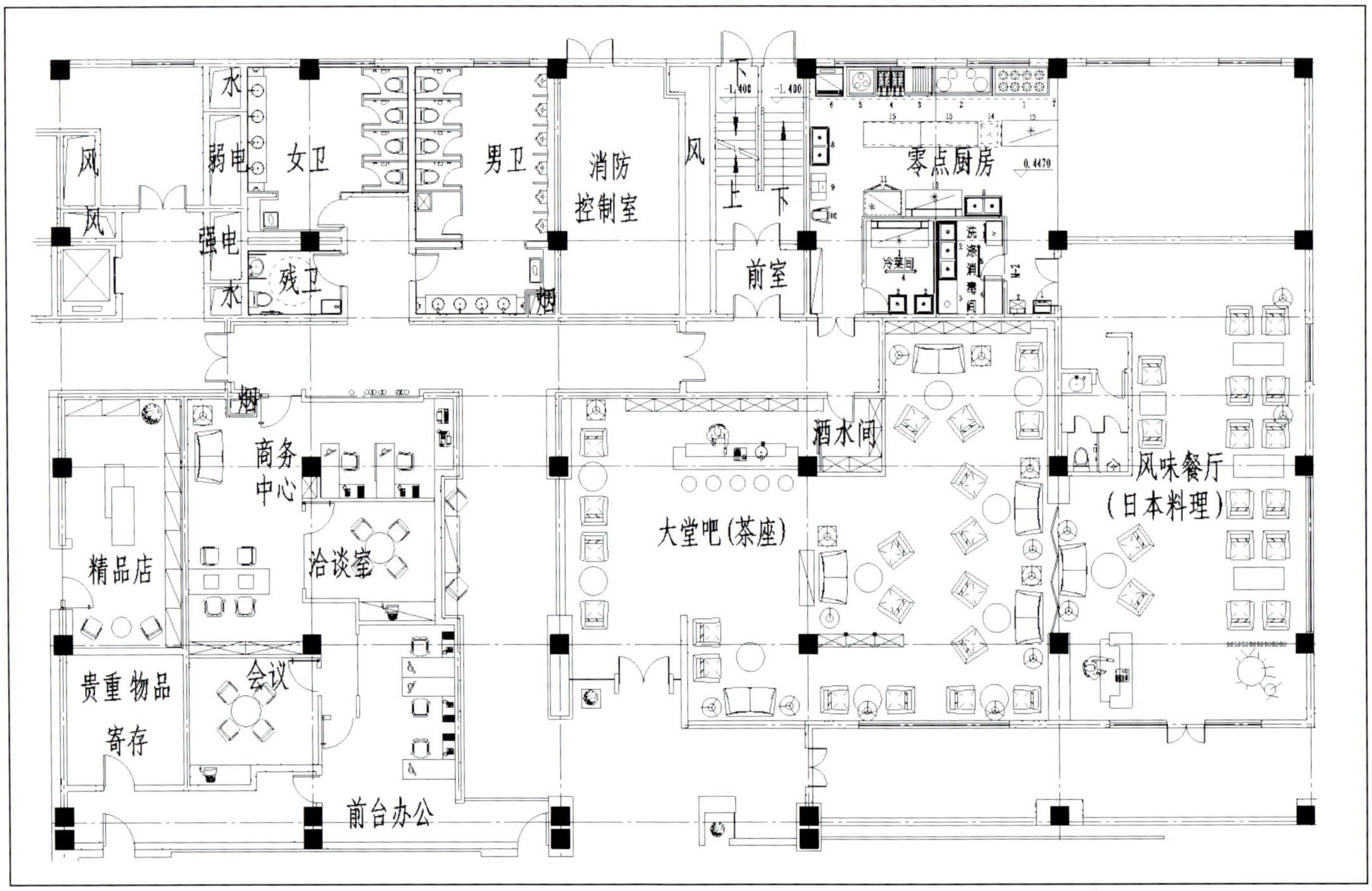

青岛海信

QINGDAO HISENSE

设计负责人：丁飞、陈刚 Architect: Ding Fei, Chen Gang

设计参加人：贝炜瑾、鲁晴宇、季婷、张颖、钱丽芬 Collaborator: Bei Weijin, Lu Qingyu, Ji Ting, Zhang Ying, Qian Lifen

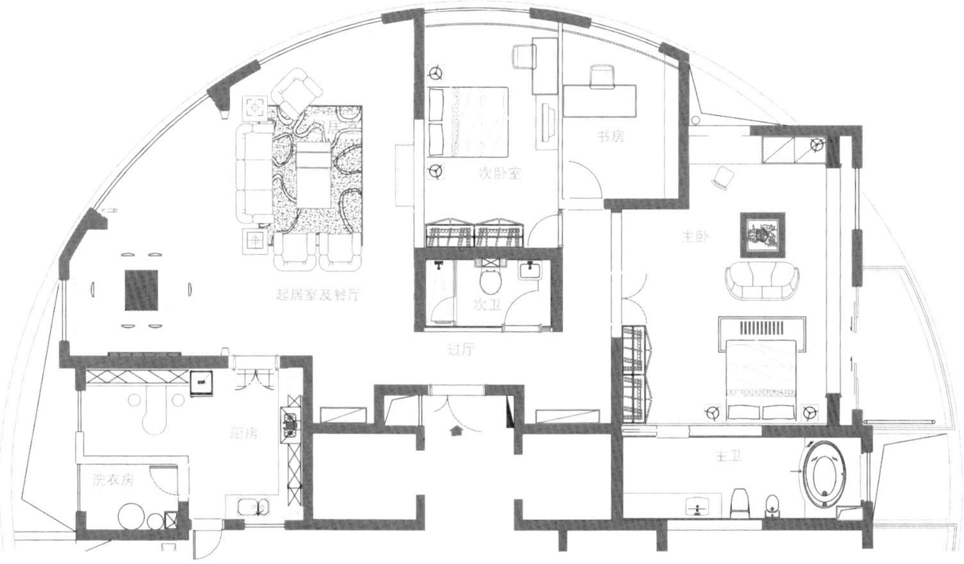

上海君临天下夜总会

SHANGHAI JUNLINTIANXIA NIGHT CLUB

设计负责人：金范九 Architect: Jin Fanjiu
设计参加人：吴纯 Collaborator: Wu Chun
建筑面积：3315㎡ Building Size: 3315m²
工程造价：1500万元 Construction Cost: RMB 15 million

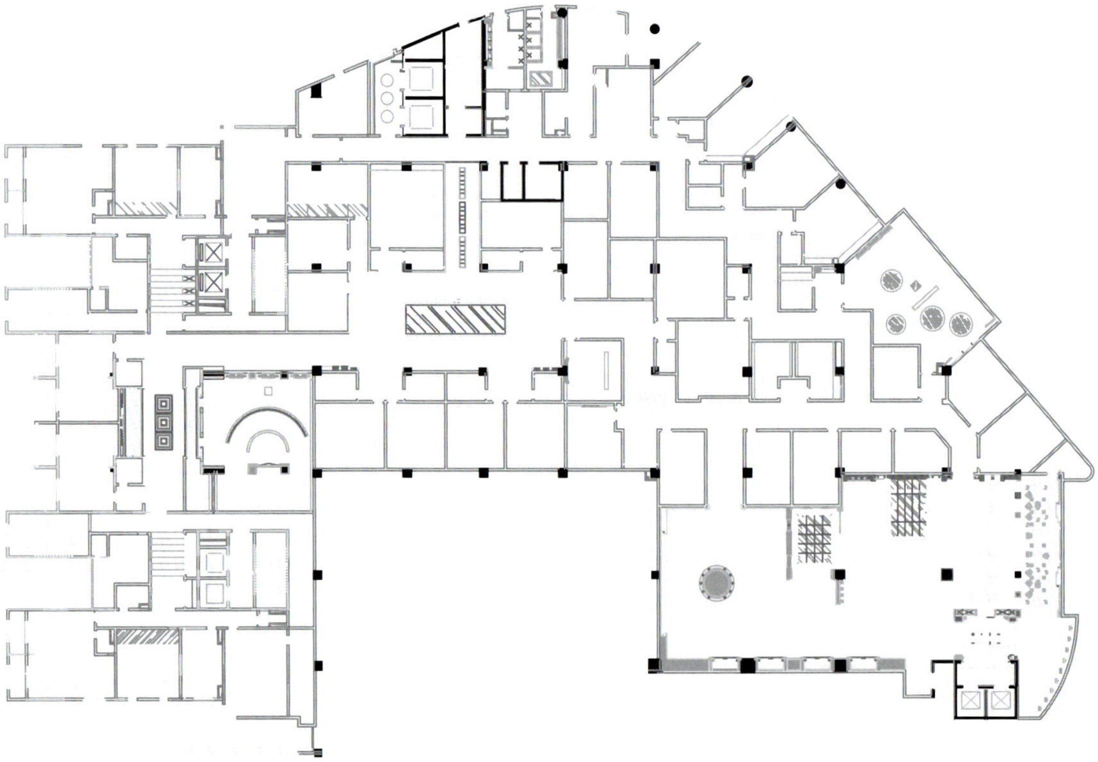

世贸滨江会所

SHIMAO RIVIERA CLUB

设计负责人：孙彦清　Architect: Sun Yanqing

设计参加人：顾洪亮、胡坤、王进　Collaborator: Gu Hongliang, Hu Kun, Wang Jin

建筑面积：6000㎡　Building Size: 6000m^2

工程造价：1000万元　Construction Cost: RMB 10 million

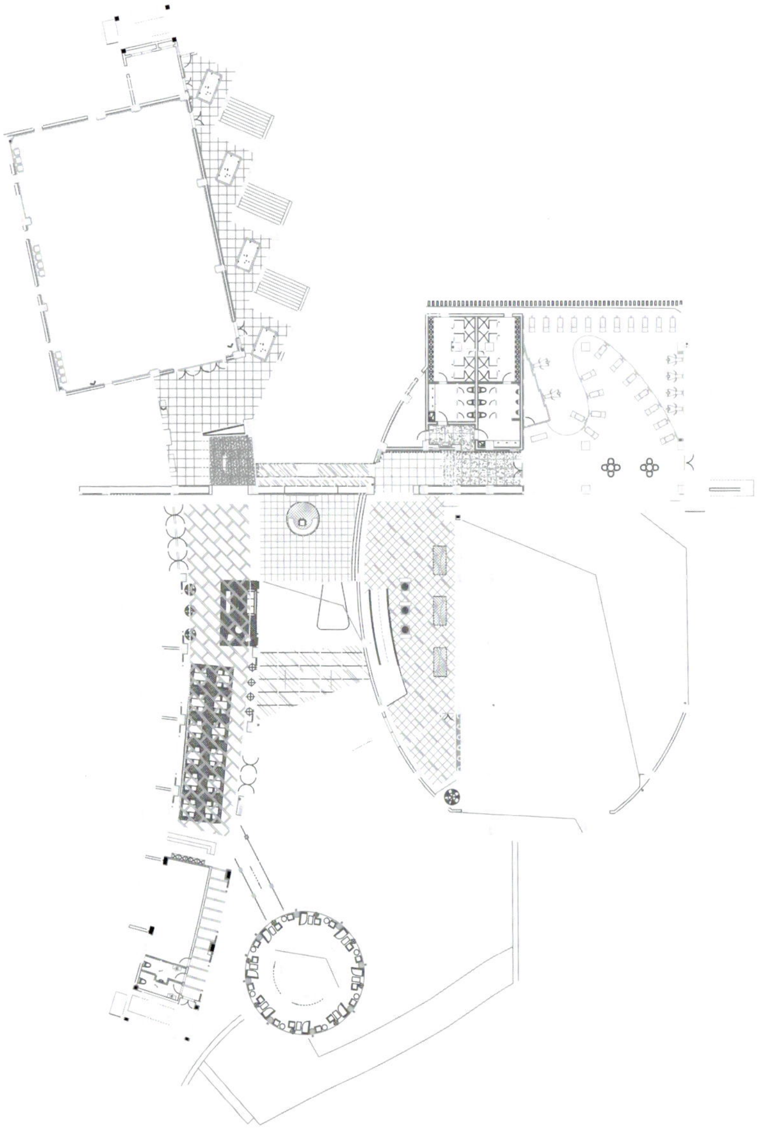

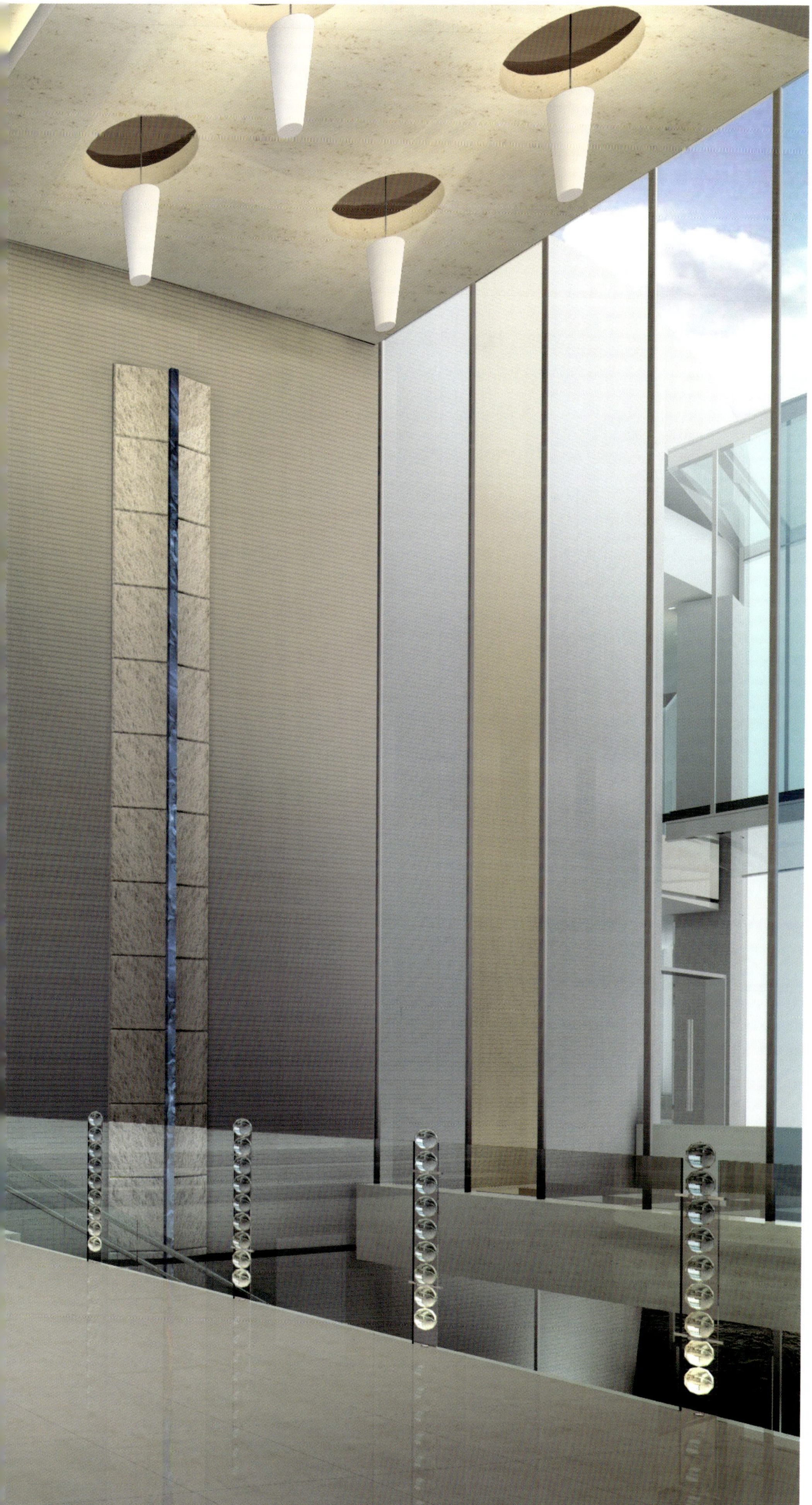

世茂集团蝶湖湾会所

BUTTERFLY BAY RESIDENCE COMMUNITY, SHIMAO GROUP

设计负责人: 范建中、沈辉　Architect: Fan Jianzhong, Shen Hui
设计参加人: 朱文奇　Collaborator: Zhu Wenqi
建筑面积: 8000m²　Building Size: 8000m²

世茂・蝶湖湾位于现代化的历史文化名城昆山中心城区南部，总占地面积约72万平方米，总建筑面积约107万平方米。容积率仅为1.5，建筑密度仅为25%，是名副其实的高绿化低密度高级住宅社区。社区由12幢沿湖面一字排开的高层公寓、数个分布于湖中的独立岛屿式的联排别墅及独栋别墅、SOHO公寓、SHOPPING MALL、星级国际酒店，三大多功能主题会所、国际双语幼儿园和学校、商业广场及台湾风情街组成。聘请国际知名规划建筑公司担纲设计，强强联合，倾情打造，感受一处桃源盛境，成就一片心灵故乡。

世茂集团蝶湖湾会所设计现代、功能齐全、气势恢弘，与周围10万平方米蓝色蝶状水系、40万平方米的湖畔公园共同构成了社区景观基础，昆山独有超大会所吸引着来自世界各地的精英共聚于此。

会所功能齐全，全面提供各种健身、娱乐、休闲配套设施，包括游泳池、男女SPA、桑拿、舞蹈室、 攀岩、 健身房、乒乓球室、桌球室、羽毛球、壁球等，营造了多元化共享空间。

Located in the south of modern and historical & cultural Kunshan inner city, with an occupation area of 720 000m², total floor space 1 070 000m², Volumetric flow rate only 1.5, and construction density only 25%, Butterfly Bay Residence Community, Shimao Group is a real high-ranked residential area with high rate of greenbelt and low density of buildings. The community is made up of 12 high-rise apartment buildings standing in a line beside the lake, a number of conjoined or detached villas on the islands in the lake, soho apartment, shopping mall, star-rated international hotel, three multi-functional theme clubs, international bilingual kindergarten and school, commercial plaza as well as Taiwan street. The famous international architectural plan and construction corporation is also invited as the designer. The win-win cooperation is to create an arcadia view, to build a spirit home.

The residence community, is with modern design, full-purposed functions and imposing appearance. With the 100000m² blue butterfly-shaped waters, and 400 000m² lakeside park as the base of the community scenery, this super residence community of Kunshan is attracting the elite from all over the world.

The residence community is with multi-functions, equipped with various facilities for body-building, entertainment and relaxation, including swimming pool, male & female SPA, sauna, dance studio, rock-climbing, fitness center, table tennis room, billiard room, badminton room, squash room and so on. It creats a diverse public area.

SHIMAO

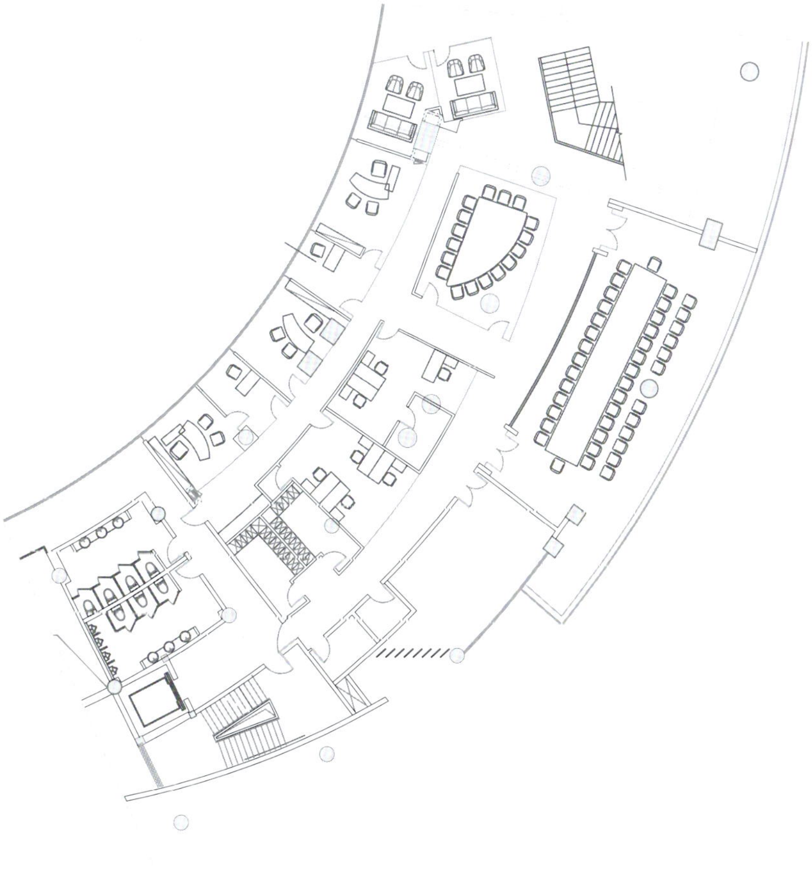

苏州电信会所

SUZHOU TELECOMMUNICATION CLUB

设计负责人: 丁飞、陈刚　Architect: Ding Fei, Chen Gang

设计参加人: 贝炜瑾、鲁晴宇、季婷、张颖、钱丽芬　Collaborator: Bei Weijin, Lu Qingyu, Ji Ting, Zhang Ying, Qian Lifen

建筑面积: 8000㎡　Building Size: $8000m^2$

工程造价: 1500万元　Construction Cost: RMB 15 million

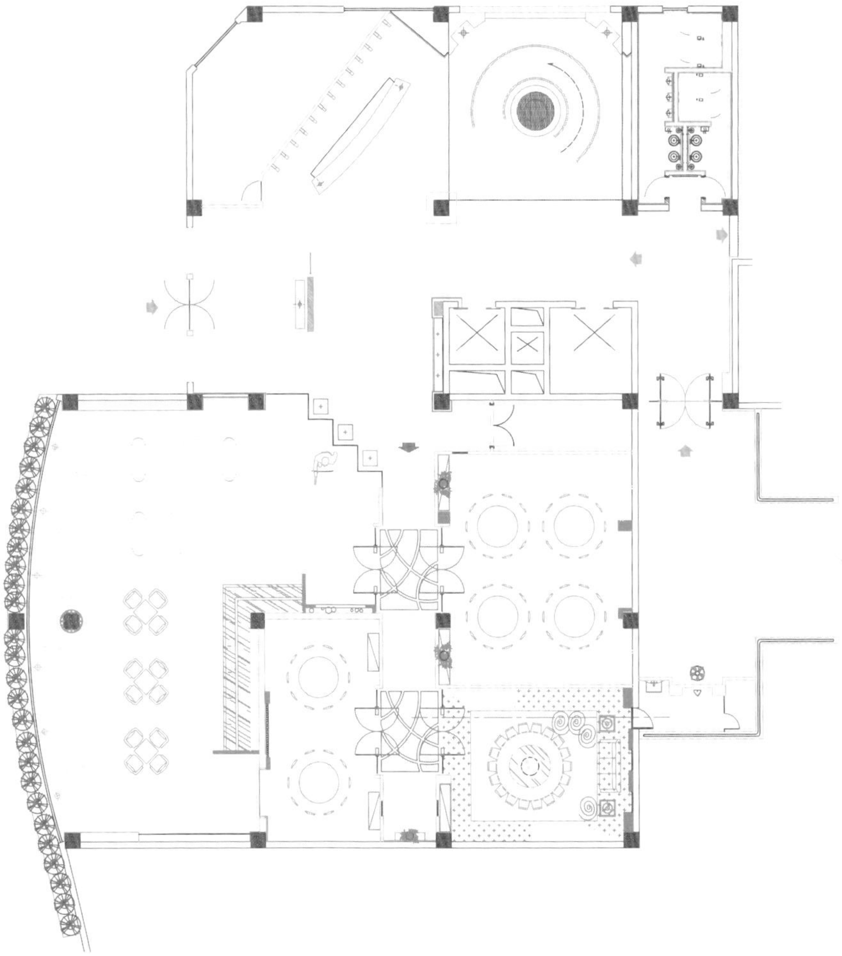

无锡新梁溪人家

XINLIANGXI RESIDENCE COMMUNITY, WUXI

设计负责人:王祎华　Architect: Wang Weihua
设计参加人:刘岩、糜佳、钱文宇　Collaborator: Liu Yan, Mi Jia, Qian Wenyu
建筑面积:7000m²　Building Size: 7000m²

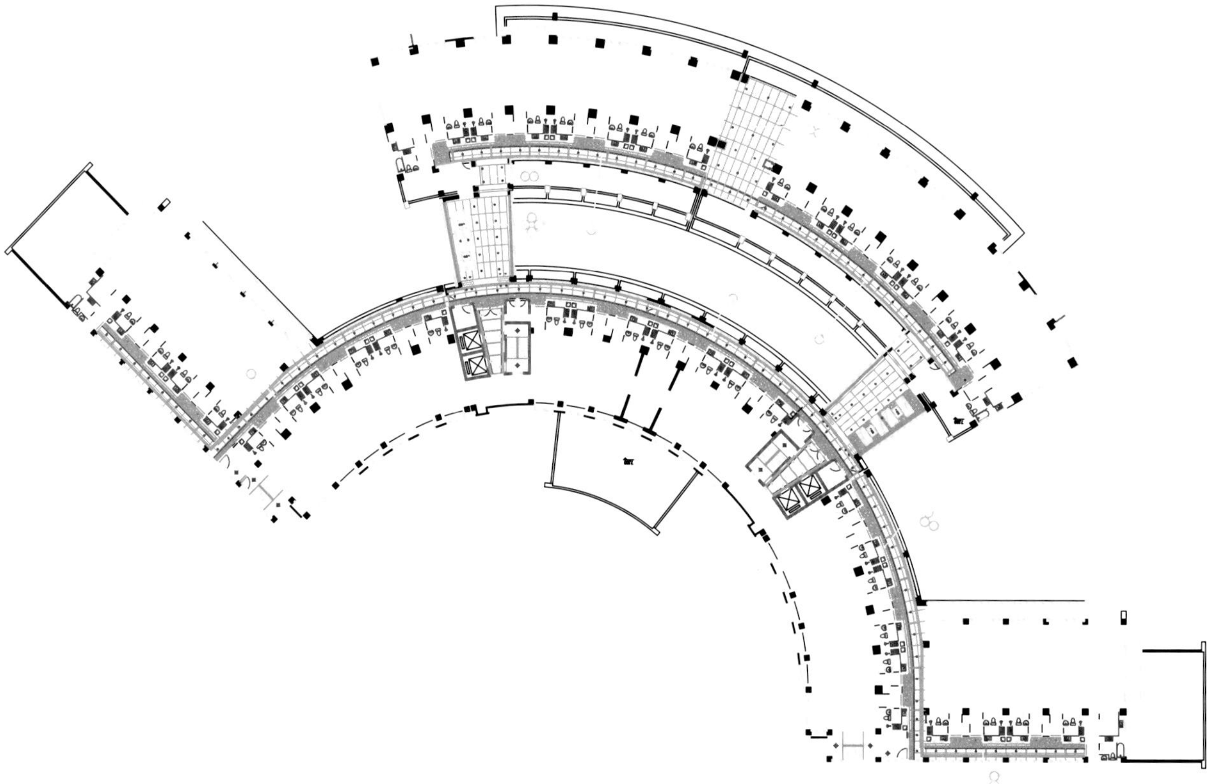

UTH
ROOMING
HOUSE

图书在版编目（CIP）数据

金螳螂设计/《金螳螂设计》编委会编．—沈阳：辽宁科学技术出版社，2008．1

ISBN 978-7-5381-4974-6

Ⅰ．金…　Ⅱ．金…　Ⅲ．室内设计—作品集—中国—现代
Ⅳ．TU238

中国版本图书馆CIP数据核字（2007）第149508号

出版发行：辽宁科学技术出版社
（地址：沈阳市和平区十一纬路29号　邮编：110003）
印 刷 者：恒美印务（广州）有限公司
经 销 者：各地新华书店
幅面尺寸：230mm×300mm
印　　张：31
插　　页：4
字　　数：150千字
印　　数：1~2500
出版时间：2008年1月第1版
印刷时间：2008年1月第1次印刷
责任编辑：符　宁　郭　健
封面设计：张士成
版式设计：张士成等
责任校对：刘　庶

书　　号：ISBN 978-7-5381-4974-6
定　　价：198.00元

联系电话：024-23284536
邮购热线：024-23284502
E-mail：tab4356@mail.lnpgc.com.cn
http://www.lnkj.com.cn